THE BIOLOGY EXPERIENCE

Laboratory Manual

Seventh Edition

Stuart J. Dearing

Ralph P. Eckerlin

Clarence C. Wolfe

Northern Virginia Community College
Annandale, Virginia

 KENDALL/HUNT PUBLISHING COMPANY
4050 Westmark Drive   Dubuque, Iowa 52002

23

CONTENTS

PREFACE

The Biology Experience Laboratory Manual was written for a one-year introductory course at the freshman college level. It is unique in that each exercise is self-contained including clearly stated objectives, adequate background information to do the exercise, a number of learning experiences and review questions. The exercises were structured for completion in approximately three hours with the possible exception of some exercises. Even those can be managed in three hours by introducing the topics prior to lab or by selectively eliminating parts. Many of the exercises incorporate some degree of overlap to provide the student with the opportunity to review previously mastered concepts and to apply them to new situations.

The microscope is introduced in the second exercise and is coupled with an introductory study of the pond. A pictorial key to the fleas is included in the introduction to taxonomy. The survey of the kingdoms is based on the five kingdom system. A copy of the Ames atlas of urine sediment is included in the exercise on physiology of the urinary and reproductive systems. Most notable among the changes in this edition is the revised sequencing of the exercises and the inclusion of many new or revised illustrations. Coverage of the phyla of the animal kingdom has been expanded into three exercises. A number of exercises have been revised extensively including: 2. The Microscope, 5. Physical Aspects of Life, 6. Cell Structure and Cell Division which combines the former exercises on Cells and Cell Reproduction and incorporates a number of electron photomicrographs with the review questions, 7. Meiosis and Genetics and 26. Pond Ecology. Ideally, the pond should be visited and studied at regular intervals during several seasons. Still, a number of insights can be gained by a single visit.

Last, but not least, the authors would like to express their appreciation to the biology faculty, staff and students of the Northern Virginia Community College for their helpful suggestions.

Safety is of paramount importance in the biology laboratory. Each student must learn and apply the safety rules as set forth below and all other safety precautions as instructed.

1. Identify the location of and learn how to operate safety equipment available in the laboratory.

2. During the first lab session learn two routes for evacuating the laboratory/building.

3. Learn the location of material safety data sheets (MSDS) and consult them for information about hazardous chemicals.

4. Report all accidents and spills immediately to your instructor.

5. Pipetting by mouth is prohibited. Use a mechanical pipetting device to transfer fluids.

6. Dispose of all hazardous chemicals in a safe and environmentally acceptable manner as instructed.

7. When dealing with body fluids (i.e. blood, saliva, urine) only work with your own or wear disposable surgical gloves and avoid skin contact.

8. Wear protective clothing and understand and use aseptic technique in handling microorganisms.

9. Disinfect all work surfaces after working with body fluids or microorganisms.

10. Discard all items contaminated with body fluids or microorganisms in designated biohazard containers.

11. Avoid eating or drinking in the laboratory and putting hands, pens, pencils, etc. in your mouth.

12. Wash your hands with soap and water before leaving the laboratory.

SCIENTIFIC INVESTIGATION

Objectives

Upon completion of this exercise you should be able to:

1. Demonstrate knowledge of the safety rules and procedures of the laboratory and identify the location of the emergency equipment and exits.
2. Name the five kingdoms of living organisms and recognize representatives from each kingdom.
3. Outline the basic concepts in the "scientific method."
4. Define the term, "hypothesis."
5. Determine the volume and dimensions of a given object or specimen using metric measurements.
6. Identify specific laboratory equipment.
7. Perform tests, examinations, make measurements on specified objects or specimens and record accurate observations.
8. Construct a hypothesis to account for an observed phenomenon.
9. Design and perform an experiment to test a given hypothesis.
10. Collect and interpret a specified set of data.
11. Explain the significance of the following factors in an experimental design:
    a. replication (repeated trials)
    b. variables
    c. control group(s)
12. Answer all of the review questions.

I. Introduction—The Kingdoms of Life

Biology is the study of life. During your study of life you will encounter a tremendous diversity of organisms of different sizes, shapes and ways of interacting with each other and with their environments. The organisms you will study throughout this course may be grouped into five broad categories (kingdoms). Two of the five kingdoms you probably already know something about—the Plant Kingdom (Plantae) and the Animal Kingdom (Animalia). In the space below write down those characteristics that you associate with plants and those that you associate with animals.

Plantae   **Animalia**

Perhaps you may have mentioned that plants are green (that is they contain the pigment chlorophyll); they are usually anchored in one place; and they can make their own food (sugar) by means of photosynthesis. For the animals you may have indicated that they move about, respond quickly to stimuli, and cannot manufacture their own food.

Located around the room are representatives of all five kingdoms. Observe each representative and briefly write down some of its characteristics in the space below:

Name of Specimen   **Characteristics**

1.

2.

3.

4.

5.

6.

7.

8.

9.

10.

Next indicate those specimens you believe are members of the same kingdom. Base your decisions on your observations and do not be afraid of making an "incorrect" choice. Use the chart below to record your choices.

Kingdoms: **Plantae**   **Animalia**   **Monera**   **Fungi**   **Protista**

Specimens:

In order to help you determine if your choices are correct, examine the brief descriptions below or consult your textbook.

Kingdom and Characteristics

MONERA—Mostly unicellular (although some form colonies or filaments); genetic material is not enclosed by a nuclear membrane; no complex membrane-bound components. Newer classification schemes subdivide Monera into two main groups or domains each of which includes more than one kingdom.
examples: bacteria including cyanobacteria (blue-green bacteria)

NOTE: Members of the remaining four kingdoms are composed of cells which have their genetic material enclosed with a nuclear membrane and they contain many complex internal membrane-bound components (organelles).

FUNGI—May be single-celled or multicellular with a body composed of branched threadlike filaments; they lack chlorophyll.

examples: yeasts, mushrooms, molds

PROTISTA—This kingdom has been termed the "trash can" kingdom by many biologists because it contains a wide variety of organisms that do not appear to belong to any of the other kingdoms. Protists do not possess any specialized tissues. They may be single-celled, colonial, filamentous, or multicellular. Some may possess chlorophyll and perform photosynthesis. In newer classification schemes Protista is subdivided into a number of groups each of which is elevated to kingdom status.

examples: algae, protozoans (ameba, paramecium)

PLANTAE—Multicellular organisms capable of photosynthesis.

examples: mosses, ferns, grasses, trees, etc.

ANIMALIA—Multicellular organisms that are unable to conduct photosynthesis and must obtain their nutrients by eating other organisms.

examples: jellyfish, worms, clams, humans, etc.

Today you were introduced briefly to the five kingdoms of life. Throughout the course you will have the opportunity to learn more about the members of each kingdom.

II. The Scientific Method

As you were observing the various organisms around the room, you may have thought of many questions to ask. Observations and question formulation represent the initial stages of the "scientific method."

Typically, scientists use an organized procedure (the scientific method) to study an organism and to solve problems. The scientific method merely represents a logical approach to solving a problem. During the laboratory portion of this course you will be exposed to a variety of biological problems, all of which may be solved by employing the scientific method. There are several basic concepts associated with the scientific method. These are:

1. Objective *observations* are performed.

2. A problem or observed phenomenon is identified. Questions are asked. Then, a statement is formulated that attempts to answer the question which in turn, helps to explain the phenomenon or solve the problem. This statement is referred to as a *hypothesis*.

3. *Experiments are planned* to test the hypothesis.

4. *Experiments are performed.*

5. *Data are collected* from the experiments. These might include measurements, color changes, behaviors, descriptions, etc.

6. The *data are interpreted* (i.e., what do the data mean?)

7. *A determination is made* in terms of whether the data support or do not support the initial hypothesis.

   a. If the data do not support the hypothesis, the hypothesis must be rejected. A rejected hypothesis may be modified and tested by additional experimentation.

   b. If the data support the hypothesis, the hypothesis is *not* rejected. This does not mean that the hypothesis is correct. It only means that current data support the hypothesis. Further experimentation may produce data that do not support the hypothesis.

A. Observations in the Laboratory.

   Making accurate objective observations is of vital importance in scientific work. This may involve the use of the laboratory and its equipment. On the next page, diagram the location of the following laboratory items: (1) exit door(s), (2) fire extinguisher, (3) eye bath, (4) instructor's/ demonstration desk, (5) sink(s), (6) trash can(s), (7) microscope cabinet(s), (8) plastic container of distilled water, (9) incubator (if present), (10) refrigerator (if present), and (11) water bath.

Hall	**Hall**	**Hall**

B. Hypothesis Formation and Experimental Design.

A hypothesis is a testable statement which suggests a possible explanation of a phenomenon or a possible solution to a particular problem. Usually a hypothesis is based on previous observations.

Suppose you were faced with the problem of which of two possible routes (the highway or backroad) to travel in order to get to class in the faster time. From previous observations you have noticed that the highway has five traffic lights while the backroad has only one traffic light, but is two miles longer. Formulate a hypothesis that suggests a possible solution to the problem of which route to take.

One possible hypothesis may be stated as follows: "By traveling the backroad to class, one may be able to complete the journey in less time than if the highway were used." Compare this hypothesis to yours. Can both hypotheses be tested? Formulate an experiment to test your hypothesis and describe the experimental procedure in the space below.

4

The hypothesis stated above may be tested by traveling each route on randomly selected days for a period of one or two weeks, making certain to leave the house at the same time each day. Record the time it takes for each journey. The data for each route then could be summed and an average time computed. If the average time for the backroad is significantly less than the average time for the highway, the hypothesis would be supported. If the backroad time was the same or greater than the highway time, the hypothesis would be rejected.

Note the word "significant" was used on comparing average times. The level of significance would have to be determined by you prior to the experiment. For example, is a two minute time difference significant? Perhaps you might feel that a difference of at least ten minutes would be required before you would choose the backroad.

In the design of this experiment certain conditions were specified. Review the experimental design and list three of the conditions specified.

(1)

(2)

(3)

One condition was that the days for travel on a particular route were *selected at random*. This prevents any bias on the part of the experimenter. Another condition specified that the departure time each day should be *constant*. This prevents the departure time from interfering with the items being tested (i.e., the two routes). A third factor was the continuation of the experiment over a period of one or two weeks. The *repeated trials* tend to minimize the chance factors such as weather conditions or traffic accidents.

In any experiment there may be many factors (*variables*) that might affect the outcome of the experiment. Therefore, experimenters attempt to control as many variables as possible. This can be done by establishing a *control group*. This group is identical to the experimental situation except for the one factor being tested. In the example above, the factor being tested was the use of the backroad. Therefore, as many conditions as possible (time of departure, randomly selected days) were kept the same during each trial run, except for the choice of road. In some experiments not all variables are subject to control. In these instances the experimenter must acknowledge the uncontrolled variables and the limitations they impose on the interpretation of the data. What uncontrolled variables were there in the previous example?

_____

_____

C. Data Collection and the Metric System.

After the hypothesis has been formulated and the design of the experiment has been completed, the next step is to perform the experiment and collect data. In the laboratory, data collection may consist of making observations, noting color changes, making counts, or determining weight, length, height or width. Measurements are made in the laboratory using the metric system and metric units. The following sections will serve to introduce you to the metric system.

1. Length—Obtain a small plastic ruler and metric tape measure (your instructor will inform you of their location). Examine the ruler first. Along both edges there are markings indicating certain units of length. Along one edge are units measured in sixteenths of an inch. On the other edge the distance between each small line represents one *millimeter* (mm).

At intervals of ten millimeters there are longer lines with a number next to each line. The distance between each number is termed a *centimeter* (cm). How many millimeters are there in one centimeter? _____ . The metric system is quite simple to use since it is based on multiples of ten. There are ten millimeters in one centimeter; there are ten centimeters in one decimeter; there are ten decimeters in one meter; and there are 1,000 meters in a kilometer. How many mm are there in one meter? _____ ($10 \times 10 \times 10 = 1,000$) Complete the following:

one meter (m) = _____ decimeters (dm)

one decimeter = _____ centimeters (cm)

one centimeter = _____ millimeter (mm)

Practice Exercises   (Note: your instructor may substitute or add other practice work)

a. Obtain a preserved specimen such as a starfish. Do all of the appendages appear to be the same length? Estimate (hypothesize) the average length of the specimen's appendages.

Hypothesis: The average length of the appendages of

specimen No. _____ is _____ cm

Measure and record the length of each appendage (carry out your measurements to one decimal place). Add all of the measurements and divide the total by the number of appendages measured. Repeat the procedure and obtain an overall average.

Length of Appendages (cm)

Specimen No. _____

Appendage length	Appendage length
_____	_____
_____	_____
_____	_____
_____	_____
_____	_____
_____	TOTAL _____
_____	AVERAGE _____

Did you measure from the CENTER of the starfish to the tip of each appendage or did you measure the appendage from the EDGE of the central body disc to the tip? A critical part of data collection is the method by which the data are collected. What difficulties could arise in the interpretation of data if the METHOD (procedure) of data collection were not given?

b. The PRESENTATION of data also is very important in scientific investigation. The use of a chart is a very effective way of presenting data. Can you think of any other types of data presentation?

In this practice exercise you will have the opportunity to obtain some data and present them by constructing a graph. First, however, hypothesize the circumference of your right wrist in centimeters.

Hypothesis: My right wrist circumference is _____ centimeters. Now, using a tape measure, determine the wrist circumference. Collect wrist data from the rest of the class or use data provided by your instructor. Construct a graph by using the vertical axis for *number of students* and the horizontal axis for *circumference*. Construct the graph in the space below:

What is the range of the data? largest = _____

smallest = _____

RANGE = _____

Which measurement is characteristic of the greatest number of students? _____

This is called the MODE. Is the mode also the average (MEAN) circumference? Calculate the average circumference and compare.

How would you present the data if you wanted to compare the distribution of wrist circumference between males and females?

2. Volume—Many laboratory experiments require the mixing of different solutions and therefore, you will be required to determine specific volumes. The apparatus that will allow you to perform this task is the GRADUATED CYLINDER.

The cylinder is made of glass or plastic. Some cylinders may rest in a red plastic base. The upper end is funnel-shaped and has a small pouring spout. Locate and examine a graduated cylinder that is about 15 centimeters in height. Along the side of the cylinder is a series of numbers ranging from one to ten. Each number rests on a long horizontal line. Each number represents a *milliliter* (ml). The shorter horizontal lines between the numbers represent one-tenth (0.1) of a milliliter. There are 1,000 ml. in one *liter*.

Locate a 100 ml beaker. Fill the beaker to the 20 ml mark. To discover the accuracy of your estimation, pour the contents of the beaker into a graduated cylinder and record the volume on the chart below. If there is a difference, which reading do you think is more accurate? You may wish to have your lab partners repeat the procedure in order to determine who is able to most accurately estimate the 20 ml volume in the beaker.

Lab Partner's Name	Beaker Reading	−	Graduated Cylinder Reading	Difference in Readings
_____	20 ml		_____	_____
_____	20 ml		_____	_____
_____	20 ml		_____	_____
_____	20 ml		_____	_____

3. Mass—Mass (the quantity of matter) is measured in grams (g) or kilograms (kg). There are 1000 grams in one kilogram. Mass may be determined by using a laboratory balance. Obtain a specimen, such as the one you previously measured for length. Estimate the mass of the specimen in grams.

Hypothesis: The specimen has a mass of _____ grams.

Test your hypothesis by using the balance. Make sure that the balance needle is on the "zero" mark. Place the specimen on the left side of the balance. Move the weight along the scaled beam until the needle returns to the "zero" position. (Note: some balances have a double beam, while those with only a single beam may require placing additional known mass units on the right pan of the balance). When the needle rests in the "zero" position, record the mass of the specimen by reading the scales and then add the sum of the extra mass units (if any) to that reading.

Specimen: _____ Mass = _____ grams

D. Data Collection and Laboratory Equipment.

In the previous section you were introduced to three laboratory tools that are used in the collection of data. The ruler, which is used to measure _____ ; the graduated cylinder, which is used to measure _____ ; and the balance, which is used to measure _____ .

In this section you will have the opportunity to discover other laboratory tools that you will use during the course. Below is a partial list. These items are located either at the demonstration desk or in one of the drawers. Find each item on the list and make note of its location within the laboratory.

Used for handling:

1. probe—all metal with blunt tip set at slight angle
2. dissecting (teasing) needle—slender metal point with wooden handle
3. forceps—two flat metal arms joined at one end

Used for cutting:

4. razor blade—single edge blade
5. scalpel—flat metal handle with blade at one end

Used for transferring:

6. pipettes—glass tubes of various sizes, tapered at one end (no stopcock present to regulate delivery of liquid)
   medicine dropper—76 mm tube with rubber bulb
   dropping pipette—100 mm tube with rubber bulb
   Mohr pipette—graduated lines along the side of the tube
   Pasteur pipette—long narrow taper at one end of tube
7. burette—very long tube with graduated lines along the side and a tapered end with a stopcock to regulate delivery of liquid
8. capillary tube—very narrow diameter (less than 1 mm) tube

Used for holding substances:

9. test tubes—varying size tubes open at one end and rounded at the closed end
10. centrifuge tube—plastic or glass tube tapered at closed end
11. fermentation tube—J-shaped tube with enlargement near opened end
12. beakers—glass containers of varying sizes with wide mouth openings and pouring spouts
13. culture dishes—circular glass dishes of varying diameters (at least 60 mm) and varying heights (at least 25 mm)
14. Petri dishes—short (20 mm or less) circular glass or plastic dishes of varying diameters
15. watch glasses—circular concave shallow glass containers of varying sizes
16. porcelain spot plates—white plates with 12 depression spots

III. Practical Exercises

A. Starch Identification.

1. At your station or at the demonstration table there is a tray containing the following equipment:

   a. one porcelain spot plate
   b. one bottle labeled starch solution
   c. one bottle labeled distilled water (H_2O)
   d. three bottles containing known chemicals and labeled:
      (1) methylene blue
      (2) iodine
      (3) phenol red

   It has been reported by certain investigators that some chemicals will undergo an easily recognizable reaction when placed in contact with another.

   The problem is to determine which known chemical (methylene blue, iodine, phenol red) may be used as a diagnostic test for the presence of starch.

   Solve this problem by using the concepts of the scientific method. On a separate sheet of paper record all hypotheses, experimental designs with controls, experimental observations and concluding statements indicating support or rejection of your hypothesis. Be prepared to submit this report at the end of the laboratory session. This report should contain the following sections:

      I. *Statement of Purpose* (what is being tested). The hypothesis should be included in this section.

      II. *Methods and Materials* (how the experiment is to be conducted). This section should contain sufficient details to enable other investigators to repeat your experiment.

      III. *Observations*. Record your data in this section.

      IV. *Conclusions*. Interpret your data and specify how they relate to the hypothesis.

2. Now that you have used the scientific method to determine a diagnostic test for the presence of starch, you will have an opportunity to apply this knowledge in the solving of a new problem. From the demonstration table obtain four bottles containing unknown solutions labeled A, B, C, and D. Using the scientific method, identify which bottle contains starch, by testing samples of each in a spot plate. Follow the format indicated in the previous section and submit the completed report to your instructor.

B. Fruit Fly Movement.

At your station or at the demonstration table there are several vials containing fruit flies. Observe the flies for one minute. Record your observations.

Select one of the following situations and formulate a hypothesis and experimental design to solve the problem.

1. the effect of light on the direction of movement of the flies
2. the effect of gravity on the direction of movement of the flies
3. the effect of temperature on the amount of movement of flies

NOTE: Lamps, an incubator, and refrigerator are available for your experimental design.

NOTE: THE FLIES MUST REMAIN IN THE BOTTLES AT ALL TIMES!

REVIEW QUESTIONS

1. Indicate the location of the eye bath in the laboratory.

2. Name the five kingdoms and give an example of an organism from each kingdom.

3. A graduated cylinder is used to measure _____ .

4. The item with two flat metal arms joined at one end and used for handling objects is called _____ .

5. Outline the basic concepts of the "scientific method."

6. Identify the term, "hypothesis."

7. If a test tube measures 12.3 centimeters in length, what is the measurement in millimeters?

8. A microscope slide measures 25.4 millimeters in width. How many centimeters would this be?

9. List two characteristics of scientific observation.

2. Stereoscopic dissecting microscope: used to view objects that are too large for the compound microscope, and for manipulating or dissecting organisms or parts of organisms.

Since both of these instruments are delicate and expensive, they must be handled with great care.

B. Handling and Cleaning. (Read this entire section *before* you obtain a microscope.) Obtaining the microscope:

1. Grasp the arm (hand grip) of the microscope firmly before sliding it out of its compartment.
2. As you are removing the microscope from the shelf, place your free hand (the hand not holding the hand grip) directly under the base of the microscope.
3. Transport the microscope in a vertical position and gently set it on the laboratory table.
4. Make note of the identification number painted on the lower portion of the arm.

Cleaning the microscope:

The lenses of the microscope are very delicate and can be scratched very easily.

1. Before cleaning the lenses, blow off any dust.
2. Use only lens paper to wipe the lens.
3. Never touch the lens with your fingers since the natural oils of your skin cause dust particles to adhere to the lens.
4. Use a soft towel to wipe other portions of the microscope.

Returning the microscope:

1. Remove all slides.
2. Clean and dry the stage of microscope.
3. Clean lenses with lens paper.
4. Place low-power objective in a vertical position.
5. Replace the dust cover.
6. Return the microscope to the shelf that has a number which matches the number on the microscope.

At this time, obtain a compound microscope from the microscope cabinet.

C. Parts and Function. (Refer to Fig. 2–1 and identify the following parts of the microscope.)

1. Base: stand which supports the microscope.
2. Arm: handle used to support the body tube and for transporting microscope.
3. Stage: flat platform which supports the slide. The opening in the center of the stage is called the aperture. On top of the stage is a mechanical stage device which holds the slide and allows it to be moved from side to side and forward and backward. The control knobs for the mechanical stage are located beneath the stage, on the right hand side. Locate the two knobs. In what direction does the upper knob move the mechanical stage?

_____

4. Substage lamp: light bulb which serves as source of illumination. Plug the microscope into a convenient outlet and turn on the lamp.
5. Substage condenser and iris diaphragm: series of lenses which concentrate light. Observe the substage condenser located just beneath the stage. Find a lever protruding from the condenser. This lever controls light intensity through the action of an iris diaphragm. Move the diaphragm lever and notice the change in the amount of light passing through the aperture.

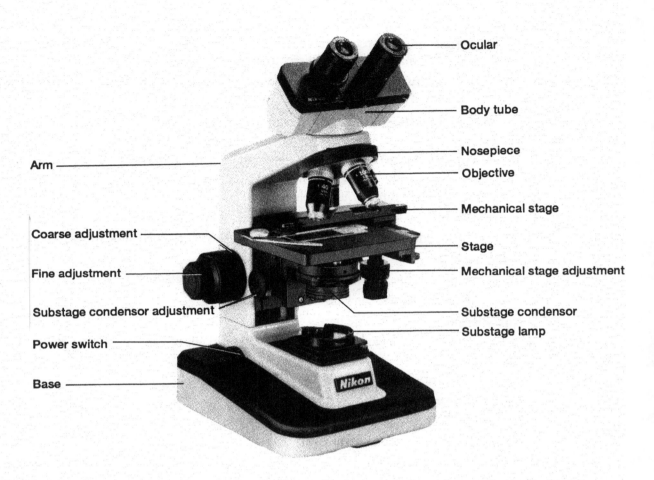

Figure 2-1. (Printed by permission IMAGE SYSTEMS, INC., Suite 304, 5430-F Lynx Lane, Columbia, MD 21044.)

6. Objective: short tube containing lenses. Its function is to form a magnified (enlarged) image. Many microscopes are equipped with four different objectives. As you examine each objective (from the side), note the number written on it. This number represents the magnifying power of the objective. Record the magnifying power for each objective listed below:

   a. Scanning-power objective: This is the shortest objective. It is very useful in obtaining an overview of the entire specimen. Its magnifying power is __4__ diameters.

   b. Low-power objective: Next to the scanning (4×) objective is another fairly short objective. It is used for more detailed inspection of a particular region of the specimen. Its magnifying power is __10__ diameters.

   c. High-power objective: The next objective is much larger than the low-power (10×) objective. It is used to make more detailed observations of a very tiny portion of the specimen. Due to its large size, however, extreme care must be employed when using the high-power objective. The magnifying power of the high-power objective is __40__ diameters.

   d. Oil-immersion objective: This is the largest objective and should only be used under special conditions as indicated by your instructor. When used, a drop of immersion oil must be placed on the coverslip of the slide (or directly on the slide) and the objective

14

must be carefully placed in the drop of oil. The oil helps retain light rays that might otherwise be lost as light passes through air. Extreme care must be taken to prevent any contact of immersion oil on the skin. Also both the slide and the objective must be cleaned thoroughly after use. The magnifying power of the oil-immersion objective is __100__ diameters.

7. Nosepiece: circular piece into which objectives are screwed; used to rotate from one objective to another.

8. Body tube: vertical or inclined structure which serves as a connection between nosepiece and ocular.

9. Oculars (eyepieces): short tubes within the body tube. They house the second set of lenses of the compound microscope and are used to magnify and view an image of the specimen.

   Examine the oculars and determine their magnifying power. The oculars magnify __10__ diameters. The 10✕ magnifying power of the oculars is identical to which of the objectives? __low power__ Normally one of the oculars contains a pointer. Look through each ocular and determine which ocular (right or left) contains the pointer. Can the pointer be rotated by simply rotating the ocular? Note that the distance between the two oculars can be shortened or lengthened.

10. Coarse adjustment knobs: large wheels on either side of the body tube. The coarse adjustment should be moved ONLY when using the scanning or low-power objective. In some microscopes the coarse adjustment raises and lowers the stage, while in others, it raises and lowers the objective. Carefully move the scanning objective into a position directly over the aperture on the stage. It should "click" into position. Now carefully rotate the coarse adjustment. Which part of the microscope moves? __stage__ . The coarse adjustment should be employed only when using which two objectives? __scanning__ and __low__ .

11. Fine adjustment knobs: smaller wheels used to sharpen final focus. Rotate the fine adjustment and observe the objective to see if it moves up or down. The movements are so fine that you can scarcely perceive them.

D. Use of the Microscope (NOTE: If you have any difficulty with any of the steps below, consult your instructor).

1. Obtain a slide labeled "letter e." Examine the slide with your unaided eye. Hold the slide so that you can read the label. With a metric ruler, measure the height of the letter to the nearest 1/10 millimeter. In figure 2–2 draw the letter (actual size) showing its orientation on the slide.

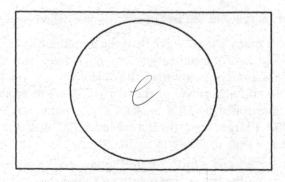

Figure 2–2.

2. Carefully swing the scanning objective in place over the aperture. Place the slide on the stage of the microscope with the label in the proper reading position and secure it in the mechanical stage.

3. Make sure that the light is on and that the iris diaphragm lever is set to allow the least amount of light through the aperture. The reduced light aids in obtaining contrast and viewing details. Move the mechanical stage in order to center the letter "e" over the aperture.

4. View the microscope from the side and slowly rotate the coarse adjustment to move the objective and stage closer together. Notice that there is still plenty of space (working distance) between the top of the slide and the objective.

5. Look into the oculars with both eyes. If you wear glasses and do not have astigmatism, you may remove them since the microscope corrects for simple nearsightedness or farsightedness. The letter "e" may not be in focus yet and may appear only as a dark blur.

6. To obtain sharp focus, slowly turn the coarse adjustment to move the objective and the stage apart. If the letter "e" was centered in the aperture, it should be visible and in sharp focus. If you fail to find the letter "e" in the field of vision, repeat the process, making sure that the "e" is centered in the aperture and that the light intensity is reduced.

7. After focusing with the coarse adjustment, use the fine adjustment to achieve a sharp clear image. Look into both oculars and at the same time move the mechanical stage to the right.

In which direction does the letter "e" appear to move? _____ . Now move the mechanical stage away from you. In which direction does the "e" appear to move?

_____ . In both cases the "e" should have appeared to move in the opposite direction of the stage movement.

8. In the first circle (field of vision) in figure 2–3, draw the letter "e." In what way does its orientation differ from your first drawing? _____ .

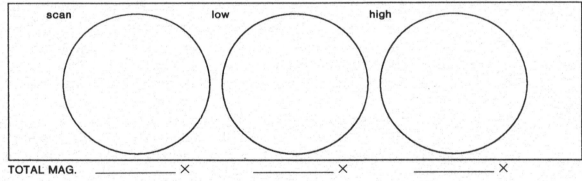

Figure 2–3.

9. The magnified image you see in the microscope when using the scanning objective is approximately how many times larger than the actual letter "e"? _____ . To calculate the degree of magnification, one must take into consideration the objective and the ocular. The magnifying power of the objective (4× for the scanning objective) is multiplied by the magnifying power of the ocular (10×) to produce the total magnification (in this case: 4× multiplied by 10× = 40×). Therefore, the image you are viewing appears forty times (40×) larger than the actual letter "e." List the total magnification "40×" in the space beneath your drawing in figure 2–3.

10. Center the letter "e" and slowly rotate the low-power objective in place. Achieve the best possible focus using the fine adjustment and then draw what you see in the middle circle of figure 2–3. Calculate the total magnification and record it in the space provided in the figure.

11. Move the mechanical stage so that the horizontal bar of the letter "e" is in the center of the field of vision. View the microscope from the side and very carefully rotate the high-power objective over the slide. Note that the objective just clears the top of the coverslip. View through the oculars and focus ONLY with the fine adjustment. Draw what you see and place it in the third circle of figure 2–3. Calculate the total magnification and record it in the space provided.

12. Since the image of the letter "e" is now enlarged approximately 400 times, the field of vision is reduced. When the low-power objective was in use (total magnification = 100×) almost all of the letter "e" could be seen. Therefore, always center in the field of vision the specific area of the specimen you wish to examine before switching to high power.

13. Switch back to the scanning objective and then remove the slide.

III. Microscopic Measurements

In the first laboratory exercise you were introduced to the metric system (the meter, decimeter, centimeter, and millimeter). These are macroscopic measurements and are used to measure objects which can be seen with the unaided eye. To measure microscopic objects, additional units of measurement are needed. These microscopic measurements include the micrometer (μm), and the nanometer (nm).

Obtain a transparent metric ruler and observe the space which denotes one millimeter. This small space (one mm) is equal to one thousand micrometers (1,000 μm). Now place the ruler on the stage of the microscope and view it under the scanning objective. Move the ruler so that the millimeter lines run across the middle (diameter) of the field of vision. What is the diameter of the field of vision

measured in millimeters? _____ The field of vision measures 4.5 mm. If there are 1,000 micrometers in one millimeter, what is the diameter of the field of vision in micrometers? _____ (4.5 × 1,000 = 4500 μm). Repeat the procedure using the low-power objective. Measure the field

of vision in millimeters and then convert to micrometers. Field of vision = _____

mm × 1000 = _____ μm. If the diameter of the field of vision under high power

is 450 μm, what is the measurement in millimeters? _____ (For this calculation simply divide 450 μm by 1,000). Knowing the approximate size of the field of vision will aid in estimating the size of specimens you examine under the microscope. Examine the prepared slide designated by your instructor, first with the scanning objective, then with the low-power objective and finally with the high-power objective. Estimate the length of the specimen. One way to accomplish this is to line up the specimen along the diameter of the field of view. If the specimen's length is about one-half of the diameter of the field of vision under high power, then the specimen measures about 225 μm (450 μm × 0.5).

The estimated length of the specimen is _____ μm.

In the space below sketch the specimen and record its name.

Your drawing may show tiny "dots" inside the specimen. These "dots" represent components within the organism that are too small to be identified even with the microscope. Cell parts are measured using a smaller metric unit of measurement: the nanometer (nm). One micrometer equals 1000 nanometers. The smallest object most light microscopes can distinguish measures about 0.22 μm.

How many nanometers does 0.22 μm equal? _____ (0.22 × 1000)

Complete the following:

one millimeter (mm) = _____ micrometers (μm)

one μm   = _____ nanometers (nm)

IV. Resolution

Magnification refers to the increase in size of the image and *resolving power* refers to the smallest distance by which two objects can be separated and still be distinguished as two objects.

Resolution (ability to distinguish detail) is largely dependent on two factors: the type of *lens* used and the wavelength of the *source of illumination*. The resolving power of the lens of the human eye is 100 μm. This means that if two objects are separated by a distance *less than* 100 μm, the human eye would not be able to distinguish the two objects, but would only perceive them as a single object. Even if the two objects were enlarged (magnified), the eye still would only detect one blurred object.

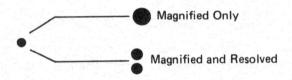

The limit of resolution of the light microscope using white light is 0.22 μm. Any object smaller than 0.22 μm cannot be seen with a light microscope. Also, objects which are separated by a distance less than 0.22 μm will appear only as a single object. The major limiting factor is the source of illumination (i.e., white light). Other microscopes have been developed which use different sources of illumination and therefore can attain greater degrees of resolution. For example, a microscope which uses ultraviolet light has a limit of resolution of 0.1 μm. The electron microscope utilizes a beam of electrons as its source of illumination and has a limit of resolution of 0.5 nm.

Write in the limit of resolution for the following in micrometers and nanometers

human eye   ____ μm   ____ nm

light microscope   ____ μm   ____ nm

electron microscope   ____ μm   ____ nm

Examine the electron photomicrograph of a chloroplast. This structure contains the green pigment chlorophyll and many other molecules. Within this complex structure light energy is captured, converted into chemical energy and used to manufacture sugar. Note the outer membrane which surrounds the entire chloroplast. Examine the inner membrane and note that it is arranged into a series of stacks and layers. This complex arrangement is vital to function of the chloroplast as you will see later in the course. Make a simple sketch of one chloroplast in the space below.

V. Depth of Focus

Obtain a prepared slide labeled "crossed threads" and view it under the scanning objective. Begin with the objective close to the stage. Move the slide so that all three colors are visible in the field of view. Each thread is composed of many slender filaments twisted around one another. If all of the filaments of all of the colors are not in clear focus, turn the coarse adjustment until sharp focus is obtained for all filaments. With the scanning power, the depth of focus is such that all three threads are in focus at the same time. Move the slide so that the intersection of the threads is in the center of the field of vision. Carefully switch to the low-power objective. Are all of the filaments of all three

colors still in sharp focus? _____. If not, use the fine adjustment to bring all of the filaments into focus. Again center the intersection of the threads. Carefully switch to high power while viewing the microscope from the side to ensure that the objective clears the top of the coverslip. Once you have successfully "clicked" the high-power objective into place, view the threads making sure that all three colors are visible, even though all may not be in clear focus. Very slowly rotate the fine adjustment and note that as the filaments of one color come into focus, the filaments of another color go out of focus. Based upon your observations of the threads, which objective provides the greatest

depth of focus? _____ . Which objective provides the

shortest depth of focus? _____ .

Parts of a specimen or object are out of focus when they are too close or too far away from the objective. Use of the fine adjustment allows one to change the distance between the specimen and the objective.

Switch back to the scanning power and remove the crossed threads slide. Obtain a prepared slide of the fruit fly, *Drosophila melanogaster*. Note the abbreviation on the label: "w.m." This stands for "whole mount" and means that the entire specimen has been mounted on the slide. View the fruit

fly under the scanning objective. Is the entire depth of the fly in sharp focus? _____ . Estimate the length of the fly. Remember the field of view under the scanning objective is 4.5 mm.

_____ . The body is composed of three regions: a head, a thorax, and an abdomen.

Switch to the low-power objective. Use the fine adjustment to view wings, legs, hairs/bristles and

eyes. Are the eyes solid or are they composed of many subdivisions (facets)? _____ .

How many pairs of legs are present? _____ . Female flies have a small projection at the end of their abdomens, while the male abdomen is round and usually much darker. The male also has a small dark comblike set of short bristles on his anterior legs. Of the two flies on your slide, is the

male on the right or the left? _____ .

Remove the slide when you are ready to proceed to the next section.

VI. The Stereoscopic (Dissecting) Microscope

Obtain a dissecting microscope or observe the one at the demonstration table. Identify the two oculars, the objective, the adjustment knob, and the stage. Place a slide labeled "the letter e" on the stage and look into the oculars. Contrast the dissecting microscope with the compound microscope in regard to:

A. the degree of magnification _____

B. the orientation of the image as seen through the ocular _____

C. the relative depth of focus _____

D. the relative working distance between the objective and the specimen _____

The dissecting microscope does not magnify as much as the compound microscope, the image is not inverted and both the depth of focus and the working distance are much greater than that of the compound microscope. Observe some of the demonstration objects using the dissecting microscope.

Examine one of your fingers using the stereoscopic microscope and focus on the fingerprint pattern. Although no two individuals have the same fingerprints, there are some basic patterns that can be identified. Figure 2–4 contains illustrations of three common patterns.

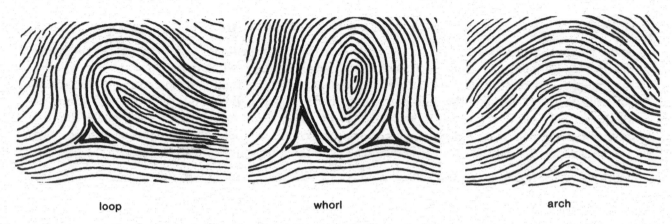

loop   whorl   arch

Figure 2–4. Fingerprint patterns.

Note the triradius, which is a triangular region where different ridge groups intersect. Which fingerprint pattern has only one triaradius? _____ . Which pattern has two triradii? _____ . Which pattern has none? _____ .
Note also the center of the pattern. If the number of ridges are counted from the center to each triradius for all 10 fingers, the median number in a large group of individuals tends to be 130. Compare your fingerprint patterns with those in figure 2–4. Do you have all three patterns represented? You may have a variation of one or more of the patterns. For example, if you have a loop, it may curve in the opposite direction.

VII. Wet Mounts

A. Technique.

When observing fresh material, a wet mount is frequently used to prevent the specimen from drying out. The basic steps in the preparation of a wet mount are:

1. Place a drop of water in the center of a flat microscope slide (some mounts may be made using a staining solution instead of water).
2. Float the material to be examined in the drop of water.
3. Place a coverslip on a 45° angle at one edge of the drop.
4. Using a dissecting needle, slowly lower the coverslip onto the specimen so that no air bubbles will be trapped.

Note: It is important to keep the preparation surrounded by water. When the water begins to evaporate, it may be replaced by adding a drop of water at one edge of the coverslip. The water will be drawn under the coverslip until the space is completely filled. Excess water may be removed by placing a small piece of paper toweling next to the opposite edge of the coverslip. The excess water will be drawn out by the toweling.

B. *Elodea*.

Prepare a wet mount of one of the small (young) leaves of the water plant, *Elodea*, and observe it under low power. The leaf is composed of two layers of rectangular cells arranged in rows. Move the slide so that the cells near the midrib are in view.

Recall the principle of depth of focus and adjust the focus knob to view first, the underlying layer of cells and then the top layer. After you have obtained the sharpest focus on one cell layer, carefully switch to high power. Notice that each cell is bounded by a rigid *cell wall*. Internal to and right up against the cell wall is a very thin *cell membrane* (7.5 nm in diameter).

Can the cell membrane be resolved with the light microscope? _____ What is the limit

of resolution of the light microscope? _____ If the width of the cell membrane is narrower than the limit of resolution of the microscope the membrane will not be resolved.

Internal to the cell membrane is a granular substance called *cytoplasm*. The cytoplasm is a highly complex arrangement of chemicals which imparts living characteristics to the cell. (Actually the cell membrane is a highly organized portion of the cytoplasm.) Suspended within the cytoplasm of *Elodea* are many small green bodies called *chloroplasts*. The chemical reactions which use light energy to manufacture sugar take place within the chloroplasts.

The process is called *photosynthesis*. The chloroplasts may appear to move in one direction around the cell. Actually it is the cytoplasm which is moving and the chloroplasts are merely being carried along. The streaming of the cytoplasm is called *cyclosis*. If you are unable to observe cyclosis, allow a few minutes for the heat of the microscope lamp to warm the cell.

By careful focusing with the fine adjustment you may observe that there is a clear area in the center of the cell in which there are no chloroplasts. This clear area is called a *vacuole* and is actually a membrane bound structure filled with water. There is another structure in the cell that might be difficult to find due to the large number of chloroplasts in the cell. This structure is the *nucleus*. The nucleus is larger and rounder than a chloroplast. It functions as the control center of the cell.

To aid in finding the nucleus, a stain can be used. The stain methylene blue has a strong affinity for the nucleic acid (genetic material) located in the nucleus. Place a drop of stain on the slide at the edge of the coverslip. At the opposite edge of the coverslip place a small piece of paper toweling and allow the water to be withdrawn by the paper toweling. Methylene blue will be drawn under the coverslip and stain the nucleus. Observe the preparation under high power. In the space below diagram one *Elodea* cell and label its parts. Estimate the length (in μm) of the cell you have drawn.

VIII. Prepared Slides

A. Tissue.

Observe a prepared slide of human blood using low- and then high-power magnification. The cells have been stained for better contrast. The most abundant cell type is the red blood cell which is round and has no nucleus. These cells average about 7μm in diameter. White blood cells are about twice the size of red blood cells and have a nucleus which has been stained blue. Can you find at least two different kinds of white blood cells based on size and the shape of the nucleus? Sketch a portion of the smear showing the red and white blood cells in the space below.

B. Organism.

Look now at the slide of the small animal the planarian. Which end is the head? Does it have legs? wings? eyes? The pharynx is the muscular structure near the center. It is used in taking in food, and is connected to a highly branched cavity (a gastrovascular cavity) from which the food is absorbed. There is no anal opening. Sketch the planarian and label the "eyes," pharynx and gastrovascular cavity. Estimate the length of the specimen.

IX. Pond Water

Using the microscope examine a wet mount of a drop or two of pond water. If some long green thread-like material (algae) is present put some of it on the slide also using the forceps. Refer to figure 2–5 to identify some of the common types of pond organisms. Try to find and sketch as many of the following as you can. Check off those that you do find.

A. General Identification.

_____ a green filamentous alga

_____ a non-green worm-like organism

_____ an organism made up of segments

_____ a ball-shaped colony of similar cells

_____ an organism with jointed legs

22

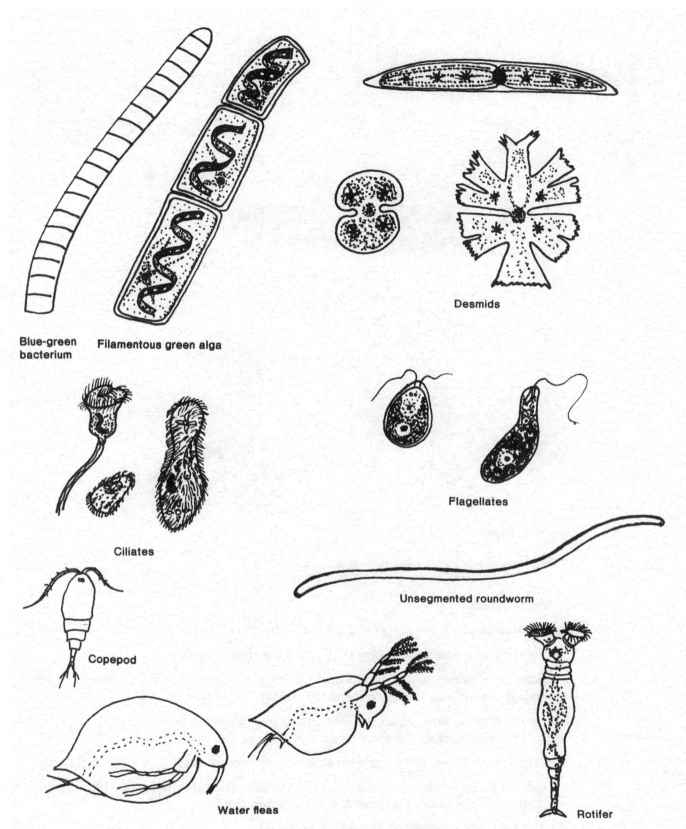

Blue-green
bacterium

Filamentous green alga

Desmids

Ciliates

Flagellates

Copepod

Unsegmented roundworm

Water fleas

Rotifer

Figure 2-5. Some pond organisms.

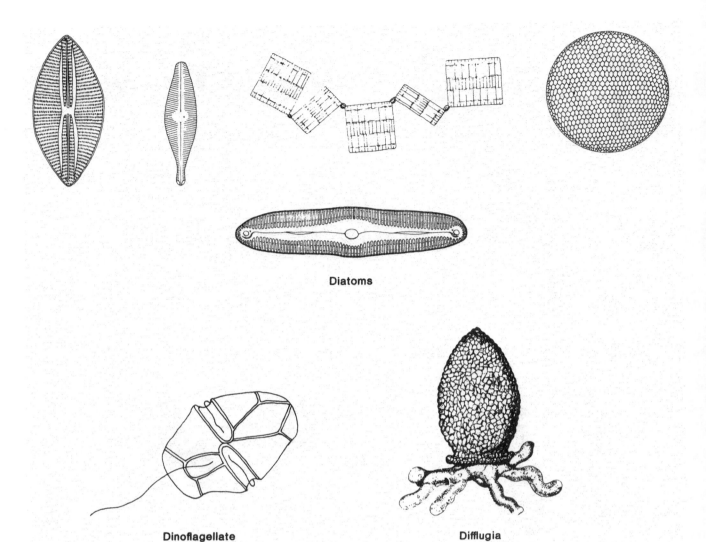

Diatoms

Dinoflagellate   Difflugia

Figure 2-5. *Continued.*

B. Specific Identification.

_____ an organism in which one half of the cell is a symmetrical copy of the other half. (desmid)

_____ an organism with fine lines (grooves) all over its surface. (diatom)

_____ a segmented organism with long spread-out antennae at one end and two smaller straight appendages at the opposite tapered end. (copepod)

_____ a segmented organism with an oval body, branching antennae extending below the head region and swimming with a jerky upward movement. (water flea)

_____ a small organism covered by a mosaic covering resembling stones in a wall. (difflugia)

_____ a slender organism with one or more rings of hair-like structures that rotate like a spoked wheel at one end and tail-like structure at the other. (rotifer)

C. The Hanging Drop Preparation.

In addition to the wet mount technique, you may wish to try the hanging drop preparation. This will allow the organisms in the sample drop to move about and will be a test of your skills with the fine adjustment control.

Hanging Drop Preparation—Place a drop of pond water on a COVERSLIP. Obtain a depression slide and carefully place vaseline around the edge of the depression. Invert the depression slide over the coverslip so that the drop is centered within the depression on the slide. Make sure that the coverslip sticks to the depression slide and then turn the slide over so that the coverslip is on the top and the drop hangs down from the coverslip into the depression. View the preparation first with the low-power objective and then switch to the high-power objective. Describe the swimming actions that you observe (e.g., jerky, whiplike movements; somersaulting movements; etc.)

In future laboratory exercises, you will have the opportunity to examine in greater detail the differences between plant and animal cells in addition to investigating the internal organization of cells.

REVIEW QUESTIONS

1. On a separate piece of paper construct a generalized diagram of a compound microscope and label all parts.

2. What is the total magnification of an object when viewed with the low-power objective (10×) and the ocular (10×)? _____

3. Complete the table below. For each item fill in the correct measurement. Place a checkmark in the appropriate column(s) (eye, light microscope, electron microscope) if the item can be resolved.

Item	Micrometers	Nanometers	Eye	Light Micro.	Electron Micro.
plasma membrane		7.5 nm			
virus	0.1 μm				
human ovum	100. μm				
bacterium		1,000. nm			

4. List two factors which determine resolving power.

_____ and _____

5. In what ways does a stereoscopic microscope differ from a compound microscope? _____

_____

6. Why is it necessary to focus the compound microscope up and down when viewing a specimen?

_____

7. Diagram an *Elodea* cell and label: cell wall, cytoplasm, vacuole, and chloroplasts.

8. Fill in the blanks indicating the approximate limit of resolution for each structure in both micrometers and nanometers.

Structure	Micrometers	Nanometers
Eye	_____	_____
Compound light microscope	_____	_____
Transmission electron microscope	_____	_____

CHEMICAL ASPECTS OF LIFE

Objectives

Upon completion of this exercise you should be able to:

1. Construct a pH scale and interpret it.
2. Measure the pH of a solution with litmus paper, pH paper, and a pH meter.
3. Define buffer, acid, base, qualitative test, and quantitative test.
4. List the four main classes of organic molecules found in cells.
5. Perform and interpret the results of a Benedict's test for reducing sugar.
6. Perform and interpret the results of the iodine test for starch.
7. Describe the product obtained by the addition of bile salts to an oil suspension.
8. Describe the specific test for a lipid and a protein.
9. Determine the carbohydrate storage substance of onion and potato.
10. Answer all review questions and complete the practice tables.

I. Introduction

Living cells are composed of numerous chemicals. Among these are four kinds of organic compounds which are important structural and functional components of cells. Numerous other organic and inorganic acids, bases and salts have important functions in living systems. For our purposes we may define an acid as a substance that can increase the H^+ (hydrogen ion) concentration of a solution. An example would be hydrochloric acid, HC1, which in a solution separates to form the H^+ and the $C1^-$ (ions). The H^+ increases the hydrogen ion concentration of the solution, so we consider hydrochloric acid an acid. Similarly, sodium hydroxide, NaOH, is a base because it is a substance that can increase the concentration of OH^- (hydroxyl ion) in solution or in other words, decrease the concentration of hydrogen ions. Table salt, NaC1 (sodium chloride), is called a salt because in solution it separates into Na^+ and $C1^-$ and increases neither the H^+ nor the OH^- concentrations. The four most important classes of organic molecules of which cells are composed are carbohydrates, lipids, proteins and nucleic acids. These chemicals are often very large and they are termed macromolecules (macro = large). We will consider their structure and learn tests to demonstrate their presence.

II. Partial Listing of Important Functional Groups

Name	Abbreviation	Structure
Alcohol Group (Hydroxyl Group)	— OH	— O — H
Carboxyl Group	— COOH	$- \overset{\overset{O}{\parallel}}{C} - OH$

26

Name	Abbreviation	Structure
Amino Group	— NH$_2$	$\begin{array}{c} H \\ \mid \\ -N-H \end{array}$
Ketone Group	— CO	$\begin{array}{c} O \\ \parallel \\ -C- \end{array}$
Aldehyde Group	— CHO	$\begin{array}{c} O \\ \parallel \\ -C-H \end{array}$
Phosphate Group	— H$_2$PO$_4$	$\begin{array}{c} OH \\ \mid \\ -O-P-OH \\ \parallel \\ O \end{array}$
Methyl Group	— CH$_3$	$\begin{array}{c} H \\ \mid \\ -C-H \\ \mid \\ H \end{array}$

III. Recognizing Elements, Covalent Bonds and Functional Groups

A. For each symbol listed below, state the name of the element, how many outer energy level elec trons are in one atom of the element and how many covalent or electrovalent bonds each elemen can form. To do this you should consult a copy of the Periodic Chart of the Elements and you biology text.

Symbol	Name of Element	Number of Outer Energy Level Electrons	Number of Bonds
C			
H			
O			
N			
P			
S			
Na			
K			
Cl			

B. For each compound listed below identify the functional groups and elements contained in one molecule and determine how many covalent bonds one atom of each of the elements has formed.

Compound	Structural Formula	Elements	Number of Bonds	Functional Groups
Methane	H | H — C — H | H			
Ethylene	H H | | H — C = C — H			

Glycine	H H O
	| | ||
	H — N — C — C — OH
	    |
	    H			

Galactose	H H OH OH H | | | | | H — C — C — C — C — C — C — H | | | | | || OH OH H H OH O			
Glycerol	H | H — C — OH | H — C — OH | H — C — OH | H			

Cysteine	H H O
	| | ||
	H — N — C — C — OH
	    |
	   H — C — H
	      |
	      S
	      |
	      H			

IV. Acids, Bases and pH

The understanding of pH is useful in chemistry, biology and many everyday life situations. pH is a measure of the hydrogen ion concentration of a solution. By definition, pH is the negative logarithm of the hydrogen ion concentration. However, when the arithmetic is performed the pH value will be a number between zero and 14. Below is a chart of the pH scale

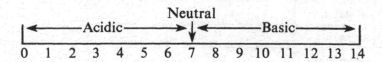

Absolutely pure water is a neutral substance. At room temperature some of the water molecules separate and 10^{-7} moles per liter of H^+ and 10^{-7} moles per liter of OH^- are formed. If a substance is added to water that increases the H^+ concentration it is an acid and it will yield a pH value of less than 7. Bases give a pH value of more than 7. Place a few drops of the substance labeled 0.1 M NaOH into a depression on the white porcelain plate called a *spot plate*. With a wax pencil label it 1.

Place a few drops of 0.1 M HCl in a depression labeled 2. Obtain two small strips of pink litmus paper and dip the end of a strip into the NaOH in depression 1 and observe if a color change occurs. If it stays pink the substance is an acid, if it turns blue the substance is a base. Discard the litmus paper strip after it is used. Repeat the test on the HCl in depression 2. Record your observations in the table below.

	Initial Color of Litmus Paper	Final Color of Litmus Paper	pH as Read from pH Paper	pH from pH Meter
NaOH	pink			
HCl	pink			

This test supplies information only about whether a substance is an acid or is not an acid. It is called a qualitative test because it determines if certain qualities are present or not present. It does not answer the question of "how much."

Now obtain two strips of pH paper. This paper has chemicals on it that will change color in the presence of acids or bases. Apply the pH paper to each of the substances in the depressions of the spot plate using a fresh pH paper for each. Compare the color of the paper to the reference colors on the pH paper bottle. Record your results to the nearest pH unit on the chart above. This is a quantitative test because it gives information on "how much."

Pour a small amount (about 15 ml) of the NaOH solution into a small beaker. In another beaker pour an equal amount of HCl. Using the pH meter at the front of the room determine the pH of the two solutions. Your instructor will show you how to use the pH meter. *Remember to rinse off the electrode with distilled water after each solution is tested, and to return the electrode to clean water when you are finished.* Record your results above.

Now obtain some carbonated soda, vinegar, urine, ammonia, and milk samples. Determine the pH of each using the two paper methods you have learned. Record your results.

	Pink Litmus Paper	pH Paper
soda		
vinegar		
urine		
ammonia		
milk		

Determine whether each is an acidic, basic or neutral substance. Indicate which test is quantitative.

Buffers are substances which when present in a solution reduce the change of pH when an acid or a base is added. Buffers are important in preventing wide shifts in the pH of blood, tissue fluid, pond water, etc. Sodium bicarbonate ($NaHCO_3$) is an effective buffer in the human body.

Label three tubes 1–3. In tube No. 1, place two ml. of distilled water. In tube No. 2, place two ml. of distilled water and one aspirin tablet (aspirin is the common name for acetyl salicylic acid, a weak acid). In the third tube place one ml. distilled water, one ml. of 5% $NaHCO_3$ solution and one aspirin. Wait until the aspirin dissolves and then determine the pH of each solution with pH paper. Pour a small volume into each of three wells of a spot plate to make it easier to wet the pH paper. Tube 1 pH = _____ ; Tube 2 pH = _____ ; Tube 3 pH = _____ .

V. Carbohydrates

Carbohydrates include organic molecules such as simple sugars (monosaccharides), double sugars (disaccharides), and polysaccharides. Glucose is a common monosaccharide with the chemical formula $C_6H_{12}O_6$. Glucose may be found in both plant and animal cells, and it is the sugar found in highest concentration in the blood of man. Maltose is a disaccharide formed by adding two glucose molecules together with the removal of a molecule of water. Maltose has the formula $C_{12}H_{22}O_{11}$.

Glucose   *Maltose*

Polysaccharides such as starch, glycogen, and cellulose are polymers formed by the repetition of numerous glucose units. Sugars are soluble in water while the polysaccharides are not very soluble in water. Both sugars and polysaccharides may serve as short- or long-term energy storage molecules in cells.

Test for Sugar

Benedict's reagent will change color after being heated in the presence of some sugars (called reducing sugars), but not others. Thus, Benedict's test is useful only in determining the presence of reducing sugars.

Label five test tubes 1–5. Place one ml. of distilled water in tube 1; one ml. of glucose solution in tube 2; one ml. of starch solution in tube 3; one ml. of onion juice in tube 4; and one ml. of potato

juice in tube 5. To each add one ml. of Benedict's reagent. What color is the reagent? _____

Place the tubes in boiling water for about three minutes, then read and record your results. A semi-quantitative estimate of the amount of sugar present can be obtained from the color of the precipitate formed.

Blue	0	No reaction
Green	+	
Yellow	+ +	Sugar present in increasing amounts
Orange	+ + +	
Dull red	+ + + +	

Record your observations.

Tube	Test Substance	Observation	Conclusion
1			
2			
3			
4			
5			

What might be a function of the onion bulb?

Test for Starch

In the presence of starch, iodine changes color from yellow-brown to a blue-black. This color change is used as the basis for a test to determine the presence of starch.

In separate wells of a spot plate place a few drops of distilled water, glucose solution, starch solution, onion juice and potato juice. To each add a drop or two of iodine reagent (I_2KI). Record your observations.

	Water	Glucose	Starch	Onion	Potato
Color Observation					
Conclusion					

In which form (starch or sugar) does the onion store most of its carbohydrates?

In which form (starch or sugar) does the potato store most of its carbohydrates?

VI. Lipids

Lipids are insoluble in water. Waxes, oils, fats, steroids and phospholipids are examples of this class of organic macromolecules. Lipids serve as long-term energy storage molecules, as structural elements in cell membranes, as insulators against heat loss and some serve as chemical messengers or hormones. They are found in every cell. Lipids can be stained with Sudan black stain.

Place two ml. of distilled water and one ml. of salad oil in each of two test tubes. Shake the tubes.

What happens? _____

Note the size of the lipid droplets.

Do two separate layers reform quickly? _____ Now add a few drops of bile salts to tube

2 and shake both again. What happens? _____

Are the droplets larger or smaller in tube 2? (Look closely) _____

This fine suspension of oil in the water is called an emulsion. Place several drops of the emulsion in a clean test tube, add a few particles of Sudan black powder and shake. Which is stained, the small

droplets of oil or the water? _____

VII. Proteins

Proteins are composed of amino acids. The amino acids link together with the loss of water to form peptide bonds. Many amino acids form a polypeptide, one or more of which function as a finished protein. There are 20 different amino acids, therefore there are 20 variations of R.

Amino acid   Dipeptide

Some examples of protein are hair, nails, feathers, scales, albumin, contractile proteins of muscle, enzymes, etc. Proteins function as structural components of membranes and microtubules, and as biological catalysts (enzymes).

Test for Protein

Biuret reagent changes color to a violet or lavender in the presence of peptide bonds. In a spot plate place a few drops of distilled water, starch solution, egg albumin solution, and skim milk in separate wells. To each add four or five drops of Biuret reagent. Observe and record results.

	Water	Starch	Egg Albumin	Skim milk
Color Observation				
Conclusion				

VIII. Unknown

Located on each lab table is a different unknown containing a reducing sugar, starch, a lipid or a protein or some combination of two of these. Determine which group is present in the unknown at your lab table and check the results with your instructor. Be sure to use a control for each test you perform.

REVIEW QUESTIONS

1. A substance that can increase the H^+ concentration of a solution is defined as a _____ .

2. Define a base.

3. Name two quantitative methods of determining pH.

   a.

   b.

4. List the four most important classes of organic compounds that make up cells. Give an example of each.

   a.

   b.

   c.

   d.

5. Benedict's test is useful to determine the presence of _____ .

6. What is the carbohydrate storage product of a potato?

   Does it differ from that of the onion?

7. Indicate whether the following characteristics best apply to sugars, starches, lipids, or proteins:

   insoluble in water—

   stains black with iodine—

   found in hair—

   stains with Sudan black—

   causes Benedict's reagent to turn green—

8. Devise a procedure to determine to which class of organic compounds an unknown substance belongs.

9. Name and draw the structure of seven functional groups.

ENZYMES

Objectives

Upon completion of this exercise you should be able to:

1. Define enzyme.
2. State the effect of pH, enzyme concentration and substrate concentration on enzyme action.
3. Determine the effects of boiling, copper ions and reducing agents on catecholase mediated reactions.
4. Cite two ways whereby enzyme activity can be detected.
5. List three properties of enzymes.
6. Answer all review questions.

I. Introduction

Enzymes are proteins that act as biological catalysts. They speed up chemical reactions that would normally occur at a much slower rate. Enzymes are a very important group of chemicals because they enable cells to perform the chemistry of life at temperatures that the cell can tolerate. Each cell contains hundreds or even thousands of different kinds of enzymes. No single enzyme can perform all the necessary catalyst functions.

Proteins have very complex three-dimensional structural configurations which are involved directly in the catalytic functioning of the enzyme. Any alteration of specific "active sites" on an enzyme's surface may affect its ability to catalyze a chemical reaction. These alterations may be produced by temperature changes, pH levels, or the presence of other molecules or ions. Some enzymes require the presence of "co-factors" in order to perform their catalytic function. For example, the enzyme catacholase requires the presence of a copper ion as a co-factor.

In the past enzymes were given descriptive names such as *pepsin* (an enzyme from the stomach) which comes from the Greek word for digestion, or *ptyalin* (an enzyme from the salivary glands) based on the Greek word for spittle or saliva. Now enzymes are named by using the root word of the substance they act upon and adding the suffix -ase. For example, sucrase—enzyme that speeds the hydrolysis of sucrose; lipase—enzyme that speeds the hydrolysis of lipids; catecholase—enzyme that speeds the oxidation of catechol. Ptyalin is now called salivary amylase because its substrate is amylose (starch). Name the enzyme that increases the rate of hydrolysis of lactose _____.

In this exercise some of the properties of enzymes will be explored.

II. Enzyme Action

A. Oxidation and Reduction.

An enzyme that occurs in potato tubers in relatively large amounts is catecholase. Peeled potatoes and bruised fruits of many kinds turn brown when exposed to the air. The color is due to melanin pigments formed by the activity of catecholase on the compound catechol.

The enzyme catalyzes the removal of two hydrogens from a molecule of catechol, with the subsequent combination of these with molecular oxygen to form water.

The red and brown pigments are composed of many quinone molecules linked together in long, branched chains. The conversion of catechol to quinone is known as oxidation, because hydrogen is removed from the catechol. Reduction, on the other hand, is the addition of hydrogen to a molecule.

The instructor will prepare a crude extract of catecholase by grinding slices of potato tuber in a blender. The extract will be filtered through glass wool in a funnel.

Work in pairs throughout the following experiments.

B. Influence of Enzyme Concentration on Rate of the Reaction.

1. Number four test tubes.
2. Fill each tube ¼ full with distilled water.
3. Next, add five drops of enzyme to tube #2, 15 drops to #3, 45 drops to #4. Then add 10 drops of catechol solution to each of the four tubes.
4. Note the time, wait five minutes and record the color intensity in each tube.

THE TUBES MUST BE AERATED BY SHAKING THEM WELL AT LEAST ONCE A MINUTE OVER THE FIVE MINUTE PERIOD.

Record the results in the table below. Indicate no color by "0," and intensity by +, + +, + + +, and + + + +.

Tube	Treatment	Color Intensity
1	No Enzyme; Catechol Only	
2	5 Drops Enzyme + Catechol	
3	15 Drops Enzyme + Catechol	
4	45 Drops Enzyme + Catechol	

What is the purpose of shaking the tubes? (check the equation preceding)

5. Plot the results obtained on the graph below.

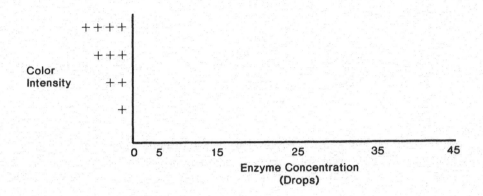

What is the effect of increasing the enzyme concentration?

C. Effect of Substrate Concentration on the Rate of Reaction.

1. Number seven test tubes.
2. Fill each tube ¼ full with distilled water.
3. Next, add one drop of catechol to tube #1, three drops to tube #2, five drops to tube #3, 10 drops to tube #4, 20 drops to tube #5, 40 drops to tube #6 and 60 drops to tube #7. Then add 10 drops of enzyme to each tube.
4. Note the time and wait 10 minutes and record the color intensity in each tube. Shake each tube once a minute while you wait.

Tube	Treatment	Color Intensity
1	1 drop catechol + 10 drops enzyme	
2	3 drops catechol + 10 drops enzyme	
3	5 drops catechol + 10 drops enzyme	
4	10 drops catechol + 10 drops enzyme	
5	20 drops catechol + 10 drops enzyme	
6	40 drops catechol + 10 drops enzyme	
7	60 drops catechol + 10 drops enzyme	

5. Plot the results obtained on the graph below.

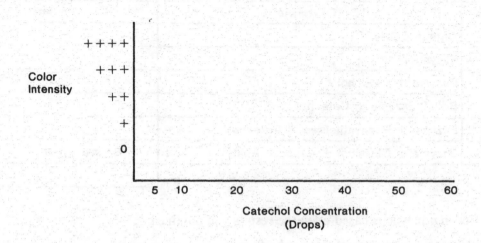

Did the color intensity level off? _____ Explain the results.

It is believed that the enzyme and substrate must meet to form an enzyme-substrate complex prior to the formation of end product. The color intensity in the latter tubes levels off because the enzyme has reached its maximum turnover rate, that is, it is meeting with substrate and catalyzing the reaction as fast as it can.

D. Effect of External Factors on the Enzyme.

Number six test tubes. Fill each tube ¼ full with distilled water. Treat each of the tubes as follows. Add everything except the catechol to each tube first. Then add the catechol to all tubes at the same time. Shake all tubes at least once a minute for five minutes.

Tube

1. 10 drops of catechol solution.
2. 10 drops of enzyme.
3. 10 drops enzyme, 10 drops catechol.
4. 10 drops enzyme, boil gently for one or two minutes, then add 10 drops of catechol.
5. 10 drops enzyme, 10 drops phenylthiourea, 10 drops catechol.
6. 10 drops enzyme, 10 drops catechol (after five minutes record the color intensity and then add 10 drops of ascorbic acid). Observe what happens and record the final intensity as the "after" observation.

Record your results in the table below.

Tube	Treatment	Color Intensity	
1	Catechol Only		
2	Enzyme Only		
3	Enzyme + Catechol		
4	Boiled Enzyme + Catechol		
5	Enzyme + Phenylthiourea + Catechol		
6	Enzyme + Catechol + Ascorbic Acid	Before	After

Answer the following questions based upon the results you obtained.

Tubes 1, 2, 3. What is the purpose of these tubes?

Tube 4. Recall that an enzyme has a complex three-dimensional shape with specific active sites on its surface. In what way did boiling the catecholase prevent its catalytic functioning?

Tube 5. Phenylthiourea is a compound which can inactivate copper. What does this suggest as to the role of copper in the functioning of catecholase?

Tube 6. Recall that the reaction in which catechol was converted to quinone was an oxidation reaction because hydrogen was removed from catechol. The agent which removed the hydrogen was oxygen. Oxygen served as an oxidizing agent in the reaction.

a. What is a reduction reaction?

b. What is the function of a reducing agent?

c. Is there any evidence that ascorbic acid is a reducing agent? Explain.

E. The Effect of pH on Enzyme Action. (Your instructor will indicate which experiment you are to perform)

1. Number six clean test tubes 1–6. Fill each tube about ¼ full with the buffered water solution indicated below. The buffer maintains the hydrogen ion concentration at or near a constant level. Add 10 drops of potato juice (the enzyme extract) to each tube. To tubes 1–5 add 10 drops of catechol. Use tube six as a color control. Observe the color intensity after five minutes and record your observations. Remember to aerate the tubes by shaking.

Tube	Treatment	Color Intensity
1	pH 3 buffer + catechol + enzyme	
2	pH 5 buffer + catechol + enzyme	
3	pH 7 buffer + catechol + enzyme	
4	pH 9 buffer + catechol + enzyme	
5	pH 11 buffer + catechol + enzyme	
6	tap water — + enzyme	

2. Plot the results on the following graph. Connect the points with a smooth curve. At which pH does the enzyme catecholase function best? _____ This is known as the optimum pH. The optimum pH may differ considerably for different enzymes.

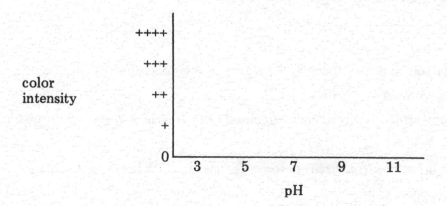

3. Amylase is an enzyme which hydrolyzes (breaks down) starch into the disaccharide maltose. This experiment tests the effect of pH on the action of amylase. Label six wells of a spot plate with the numbers "3", "5", "7", "9", "11" and "control." Place four drops of buffered solution in each well as indicated on the chart below. The buffer maintains the hydrogen ion concentration at or near a constant level. Add one drop of a 1% starch solution to each well. Then add one drop of I_2KI to each well. Record the appearance of each well on the chart below. Next add five drops of 5% amylase solution to each well except the control well. Stir each well with a separate toothpick. After 10 minutes record the appearance of each well on the chart below.

Well	Treatment			First Observation	Treatment	Second Observation
	Buffer 4 drops	Substrate 1 drop	Indicator 1 drop	(Color)	Enzyme	(Color)
3	pH3	1% starch	I_2KI		5 drops 5% amylase	
5	pH 5	1% starch	I_2KI		5 drops 5% amylase	
7	pH 7	1% starch	I_2KI		5 drops 5% amylase	
9	pH 9	1% starch	I_2KI		5 drops 5% amylase	
11	pH 11	1% starch	I_2KI		5 drops 5% amylase	
Control	Tap water	1% starch	I_2KI		None	

At which pH did the amylase work the fastest? _____

At which pH(s) did the amylase fail to work? _____

Based on your knowledge of enzyme active sites and three-dimensional structure, explain the effect of pH change on enzyme performance.

REVIEW QUESTIONS

1. In what class of organic compounds do enzymes belong? _____

2. Suggest a name for the enzyme that catalyzes the hydrolysis of:

   cellulose _____

   maltose _____

3. What was the source of the catecholase enzyme used in this exercise? _____

4. The iodine test is used to test for _____ .

5. Catechol is converted to quinone by the removal of hydrogens which combine with _____ _____ to form water.

6. The catechol is (reduced/oxidized) to form quinone. _____

7–9. List three properties of enzymes:

   _____

   _____

   _____

10. The pH of the human stomach is about 1.5. Is this acidic or basic? _____ .

   What do you suppose is the pH optimum for human stomach enzyme activity? _____

PHYSICAL ASPECTS OF LIFE

Objectives

1. Provide a definition for diffusion, osmosis, dialysis, solution, solute, and solvent.
2. Perform and describe experiments that demonstrate the diffusion of substances in air, water, and a gel.
3. List at least two factors which influence the rate of diffusion and apply this knowledge to the interpretation of experimental data.
4. State the importance of osmosis and diffusion to living organisms.
5. Perform and describe the glucose-starch dialysis experiment and interpret the results.
6. Perform and describe the osmosis experiments.
7. Answer all review questions in the exercise.

I. Introduction

The molecules of matter are in constant, rapid, random motion. Because molecules are moving they possess kinetic energy. The more rapidly molecules move the greater the kinetic energy of the molecules. Thus, when water is heated, water molecules move more rapidly. When the water molecules are restricted to a closed container (e.g., a pressure cooker or a car radiator) they exert a pressure on the walls of the container. When water is in an open container, the molecules tend to move farther apart (e.g., water evaporating from the surface of the body). This inherent movement of molecules is the basis of several biologically important physical processes: diffusion, osmosis and dialysis.

From the standpoint of a living organism, diffusion and osmosis are extremely important. These are two of the processes involved in getting needed materials into cells and waste products out of cells.

Since some of the experiments require a full hour or more to run, you should skip to sections IVA and B and prepare the glucose solutions, sections VA and B and set up the osmosis experiments and section VI and set up the dialysis experiment. Allow each experiment to run for the indicated amount of time. During the waiting period proceed through the remainder of the exercise beginning with section II.

II. Evidence for Molecular Motion

Prepare a wet mount of a yeast suspension (or of India ink) and examine it using the high dry objective. Describe the type of movement exhibited by the yeast cells.

_____

_____

The yeast cells should appear to jiggle back and forth. The random movement exhibited by microscopic particles (yeast cells, India ink particles, etc.) suspended in water is called Brownian movement. Molecules of the water in which the yeast cells are suspended are in constant random motion. As water molecules bump into the larger yeast cells some of their kinetic energy is imparted to the suspended cells. Next, examine the yeast suspension using the oil immersion objective and describe the movement of the oil droplets located in the yeast cell cytoplasm. Use lens paper to clean the oil from the 100× objective when you finish your observations.

_____

_____

A vacuole can be readily distinguished in the center of each yeast cell by carefully focusing up and down with the fine adjustment knob. Also, budding yeast cells should be visible. Budding is an asexual form of reproduction.

III. Diffusion

The molecules of a substance, whether gas, liquid or solid, tend to move from a region of higher concentration of that substance to a region of lower concentration of that substance until the molecules are dispersed evenly throughout the space available to them. This movement, called diffusion, is a direct result of the kinetic energy of the molecules.

   A. Diffusion of Gases (to be demonstrated by instructor).

   The diffusion of gases through air can be demonstrated with a length of glass tubing attached to a ring stand. Prepare small cotton plugs to fit the ends of the glass tube. Next, soak one cotton plug with hydrochloric acid (HCl) and the other with ammonium hydroxide (NH_4OH). (Care should be exercised in handling the solutions.) Place the cotton plugs into opposite ends of the glass tubing simultaneously (see Figure 5–1) and observe what happens.

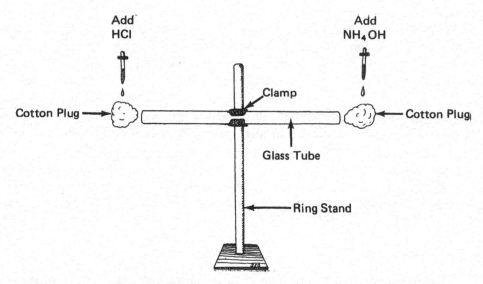

Figure 5–1. Apparatus for diffusion of gases.

The ammonium hydroxide reacts with hydrochloric acid to form a white precipitate, ammonium chloride (NH_4Cl), plus water.

$$NH_4OH + HCl \rightarrow NH_4Cl + H_2O$$

Indicate on Figure 5–1 the location of the NH_4Cl and H_2O which are formed. The reaction occurs closer to which end of the test tube? _____ Actually, the white ring forms when the ammonium (NH_4^+) and chloride (Cl^-) ions meet. The formula weight of the ammonium ion is 18.0 and that of the chloride is 35.5. Based on the location of the product, NH_4Cl, and the weights of the ammonium and chloride reactants, formulate a statement which describes a factor which influences the rate of diffusion.

B. Diffusion of Substances in Water and in a Gel.

The molecules of a liquid are farther apart than those in a gel. Will this have any effect on the rate of diffusion? Formulate a tentative hypothesis concerning the relative rates of diffusion of a substance in a liquid and in a gel. _____

_____

Test your hypothesis by performing the following experiment. Fill a petri dish with water. Place the dish on the table and let it sit until there is no water movement. Obtain a petri dish containing agar. Also get two crystals of potassium permanganate and place one crystal in the center of each dish. Observe the two dishes for five minutes and record the results.

In which dish has diffusion been more rapid? _____

IV. Solutions and Concentration

A solution is a homogeneous mixture consisting of a solute dissolved in a solvent. In biology, water is the most important solvent. One way of describing the concentration of a solution is per cent of solute by weight (per cent of the total solution weight contributed by the solute).

A. Weight of Water

Obtain, weigh and record the weight of an empty 100 ml graduated cylinder. Next, fill the graduated cylinder with distilled water to the 100 ml mark and determine and record the combined weight. Finally, determine the weight of 1 ml of water to the nearest tenth of a gram.

Wt. 1 ml H_2O = (Wt. graduated cylinder plus H_2O—Wt. graduated cylinder)/100

_____ g.   _____ g.   _____ g.

B. Preparation of Solutions

1. Prepare a 1% sucrose solution by weighing and dissolving 1 g. of sucrose in 99 ml of distilled water.

2. Prepare a 3% sucrose solution by weighing and dissolving 3 g. of sucrose in 97 ml of water.

Which of the above solutions has the greater concentration of solute? _____

Which of the above solutions has the greater concentration of water? _____

V. Osmosis

The diffusion of a solvent (water) across a semipermeable membrane is called osmosis.

A. Obtain 6 large or 8 small raisins and subdivide them into two groups based on size. Determine and record the collective weight of the raisins in each group to the nearest tenth of a gram. Place the raisins of one group (Group 1) in a small beaker and cover them with water. At the end of two hours determine and record the collective weight of the raisins for each group. Before weighing the raisins in the water covered group roll them on a paper towel to remove excess water.

Group	Initial Wt.	Final Wt.	Change in Wt.
1 (in water)	_____	_____	_____
2 (not in water)	_____	_____	_____

Explain the results based on the principle of osmosis. _____

beaker contents ———   ——— sac contents

Figure 5–2. Sac-Beaker arrangement.

B. Sucrose Sacs in Water

Obtain four 25 cm long sections of dialysis tubing and twist and tie off one end of each with thread to form four bags. Pour 40 ml of 1% sucrose solution into each of two bags. For each bag, remove the excess air from the open end, twist and tie off both ends together to form a U-shaped sac. Refer to Figure 5-2. Similarly, prepare two additional sacs, each containing 40 ml of 3% sucrose. Determine the weight of each sac to the nearest tenth of a gram and record the data in the space provided. (Note: make sure the balance is set at zero before each weighing.) Obtain four 250 ml beakers. Add 50 ml of distilled water to two of the beakers. To the third beaker add 50 ml of 1% sucrose and to the fourth beaker add 50 ml of 3% sucrose. The latter two beakers will serve as controls. Which sac will be placed in the 1% sucrose beaker to serve as one control? _____ Prepare a second control by placing a 3% sucrose sac in the 3% sucrose beaker. For the experimental groups, place each remaining sac into separate distilled water beakers. Record the start time. At the end of one hour, remove each sac, blot it gently to remove excess water, weigh it, and record the weight.

Sac contents	Initial Wt.	Final Wt.	Change in Wt.
1% sucrose **control**	_____	_____	_____
1% sucrose **experimental**	_____	_____	_____
3% sucrose **control**	_____	_____	_____
3% sucrose **experimental**	_____	_____	_____

Explain the results using the principle of osmosis. _____

_____

Predict what would have happened if the sac containing the 1% sucrose solution had been placed in the beaker with the 3% sucrose solution. _____

Hyper = over or more than   Hypo = under or less than

When two solutions are compared, the solution having the greater solute concentration is termed *hypertonic;* the solution with the lesser solute concentration is termed *hypotonic.* For each pair below, place a check in front of the hypertonic solution.

_____ Distilled water   _____ 3% glucose

_____ 1% glucose   _____ Distilled water

_____ 1% glucose   _____ 3% glucose

_____ 3% glucose   _____ 5% glucose

(Note: Whether any given solution is hypotonic or hypertonic depends on the solute concentration of the solution with which it is compared. Also, the hypotonic solution contains a higher concentration of water than the hypertonic solution.)

C. Plasmolysis

Prepare a wet mount of one of the small leaves of *Elodea* or of red onion epidermis and examine the preparation under the microscope to make sure the cells are normal and healthy. Draw one or two cells in the space provided below. Then, place a piece of filter paper on one side of the coverslip to absorb water. As the water is absorbed replace it with salt solution by putting a drop of 5% NaCl next to the coverslip at the side opposite the filter paper. Observe the cells after the salt water reaches them and draw one or two cells below.

Before   **After**

What happens to the protoplast (living substance) of each cell? Explain why!

_____

_____

Is the cell wall affected? _____

Define plasmolysis! _____

_____

In plasmolysis, water moves out of the cell protoplast by osmosis resulting in the protoplast shriveling up and moving away from the rigid cell wall.

VI. Dialysis

The separation of small molecules from large molecules based on the differential permeability of a semipermeable membrane is termed dialysis.

Obtain a 25 cm piece of dialysis tubing and twist and tie off one end with thread. Place 40 ml of a glucose-starch mixture in the bag and twist and tie off both ends together to form a U-shaped sac. Put the sac in a 250 ml beaker with 50 ml of distilled water and let it stand for one hour. See Figure 5–2. After one hour remove, wash and blot the sac dry. Be sure to retain the contents of the beaker. Break the sac open into a second beaker labeled sac contents and mix the contents around. Transfer 1 ml of the sac contents into a clean test tube and 1 ml of the beaker contents into a separate clean test tube. Test for the presence of starch using iodine solution (I_2KI). Record the results as positive or negative below.

Transfer 1 ml of the sac contents into a clean test tube and 1 ml of the beaker contents into a separate clean test tube. Test for the presence of glucose (a reducing sugar) using Benedict's reagent. Record the results below.

Test	Sac Contents	Beaker Contents
Iodine test for starch	_____	_____
Benedict's test	_____	_____

Did the starch molecules pass through the membrane? _____

Did the glucose molecules pass through the membrane? _____

Are the molecules of starch or glucose larger? _____

Would you expect a 1% glucose solution or a 1% starch solution to contain more particles? Explain your answer.

_____

Since the small glucose molecule can pass through the sac membrane, both the sac and beaker contents should test positive for reducing sugar. Because the starch molecule is too large to pass through the membrane, only the sac contents should test positive for starch.

REVIEW QUESTIONS

1. Microscopic cells or particles suspended in water exhibit a type of movement known as

   _____ .

2. The yeast cells in a yeast suspension appear to jiggle back and forth because they are bombarded by the _____ of water which are in constant, rapid, random motion.

3. The movement of the molecules of a gas, liquid, or solid from a region of higher concentration of that substance to a region of lesser concentration is termed _____ .

4. Two factors that influence the rate of diffusion are _____ and _____ .

5. _____ is the diffusion of a solvent across a semipermeable membrane.

6. In _____ , water moves out of a cell by osmosis causing the protoplast to shrivel up.

7. In _____ , small molecules can be separated from large molecules based on the differential permeability of a semipermeable membrane.

8. The process of dialysis can be used in treating kidney patients to separate urea from proteins because the urea is _____ to pass through the dialysis tubing.
   (too large or small enough)

9. Which type of solution (hypotonic or hypertonic) is formed in the soil environment around root cells when a lawn is over fertilized? _____

10. Would the root cells lose or gain water? _____

11. Why is it common practice to place lettuce and celery in water before serving them?

CELL STRUCTURE AND CELL DIVISION

Objectives

Upon completion of this exercise you should be able to:

1. Distinguish between a eukaryotic cell and a prokaryotic cell by giving examples of each type and by identifying electron photomicrographs of each type.

2. Prepare wet mounts of onion, potato, red pepper, and epithelial cheek cells and identify the structures of each.

3. Recognize from micrographs, drawings, wet mounts or prepared slides the following components of the cell and give the functions of each: cell membrane, nucleus, chloroplasts, leucoplasts, vacuole, cell wall, mitochondria, endoplasmic reticulum, ribosomes, Golgi apparatus, lysosomes, microtubules, cilia, flagella, and centrioles.

4. Differentiate between a plant cell and an animal cell in micrographs, wet mounts or prepared slides.

5. Recognize the various phases of a cell cycle from drawings, microscope slides or models.

6. From memory make drawings depicting the phases of a cell cycle and write a brief description of the events which occur in each phase.

7. Distinguish between cell division in plants and animals by means of drawings, slides and models.

8. Answer all of the questions in the review section.

I. Introduction

In the previous exercise on the microscope you had the opportunity to prepare a wet mount of a leaf of the water plant, *Elodea*. By using the microscope you were able to observe that the *Elodea* leaf was composed of many small units called *cells*. All living organisms are composed of cells. Most organisms, including you, exist at one time in their development as a single cell. Some organisms spend their entire life as a single cell, while other organisms develop from a single cell into a complex organism made up of billions of cells.

The shape and size of cells vary tremendously. A single nerve cell in your leg may measure up to one meter in length, while one of your red blood cells may be only seven micrometers in diameter. Some cells exist independently by being able to perform all of the basic functions of life, while other cells are highly specialized and are limited to the performance of one or two functions. Even with all of this diversity most cells possess a similar structural organization. In this exercise you will have the opportunity to examine various cells and to identify some of the basic components of these cells. The mechanism by which cells reproduce will also be examined.

II. Cells

A. Eukaryotic and Prokaryotic Cells

All cells are composed of a complex arrangement of molecules and ions known as *cytoplasm*. The cytoplasm is the living substance of the cell. The molecules on the outer edge of the cytoplasm are arranged in a specific order forming the *cell membrane*. The cell membrane forms a boundary between the cell and its environment and helps to regulate the passage of materials into and out of the cell. Another feature common to all cells is the presence of *genetic material, DNA*. DNA specifies the kinds of proteins cells can manufacture, and, therefore, determines the basic activities of cells.

Within the cells of many organisms the genetic material is located in a *nucleus* separated from the cytoplasm by a *nuclear envelope* (porous double membrane). Such cells are termed *eukaryotic* (true nucleus). Cells in which the genetic material is not separated from the cytoplasm are *prokaryotic*. True bacteria and blue green bacteria have prokaryotic cells. All other cellular organisms (animals, fungi, plants and protists) have eukaryotic cells. In comparison to prokaryotic cells, eukaryotic cells are larger and contain a number of membrane bound *organelles*. Organelles are compartments in the cytoplasm that perform specialized functions such as photosynthesis, ATP synthesis or packaging of secretions.

Observe the electron photomicrographs shown below and label each cell as eukaryotic or prokaryotic. Also, with the aid of your textbook or instructor identify the cell features shown at "a" through "f".

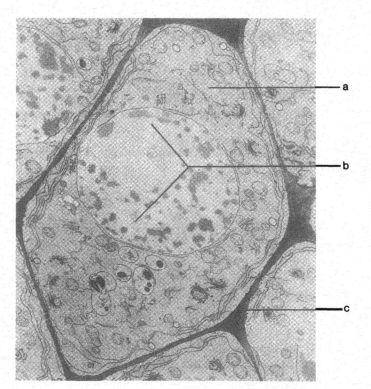

Courtesy of H. H. Mollenhauer

Courtesy of C. F. Robinow

Cell type _____

Cell type _____

B. Eukaryotic Cells Under the Light Microscope
  1. Plant Cells
    a. *Elodea*
       Diagram and label from memory an *Elodea* cell.

       Your diagram should show a rectangular cell with a relatively thick *cell wall,* a *thin cell membrane,* a *vacuole, cytoplasm,* a *nucleus* and some *chloroplast* organelles. If necessary, prepare a fresh wet mount of *Elodea* and observe its basic structure.
    b. Onion (*Allium*)
       An onion is composed of layers of scales (modified leaves) that can be easily separated. The inner (concave) surface of each scale is covered by a thin layer of cells called the *epidermis.* This layer may be removed by cracking the scale and removing the epiderimis with forceps. Prepare a wet mount of onion epidermis and compare onion cells to *Elodea* cells. Observe the preparation under low power and then switch to high power. Diagram an onion cell and label cell wall, nucleus, nucleoli and cytoplasm.

       If you have difficulty identifying structures, add a drop or two of iodine solution to the edge of the coverslip. What structures are common to both the *Elodea* cell and the onion cell?

       _____

       _____

       _____

       Since both cells are eukaryotic, they should possess a cell membrane, cytoplasm, and a nucleus. Since both cells are plant cells they should possess a cell wall and a relatively large vacuole. Note that the onion cell lacks chloroplasts.
    c. Potato (*Solanum*)
       Next take a thin slice of tissue from the interior of a potato and prepare a wet mount. View it under low power, and then switch to high power. Add a drop of iodine solution to the edge of the coverslip. Record your observations of the tissue after contact with the iodine. Sketch a small section and label a leucoplast.

       _____

You probably observed that small oval blue-black structures appeared within the cell. These structures are specialized organelles called *leucoplasts*. They function as starch storage units within the potato. The blue-black color was produced when the iodine combined with the starch.

Both leucoplasts and chloroplasts are examples of one group of organelles called *plastids*. Plastids are found only in plants and protists. Chloroplasts function in the process

of photosynthesis. Leucoplasts function in _____ storage.

d. Red Pepper (*Capsicum*)

Plants may contain other plastids called *chromoplasts*. Chromoplasts contain pigments which give the plant some of its coloration. Carotenoid pigments are the chief pigments found in chromoplasts. These pigments are largely responsible for the yellow to red coloration in plants. In many plants the presence of the carotenoids is hidden by the presence of large amounts of chlorophyll in the chloroplasts. Under certain conditions the chlorophyll breaks down and the carotenoids become visible. For example, when a patch of grass is covered for several days, the green pigment fades and the grass appears yellow.

There are two major groups of yellow to orange carotenoids, the carotenes and xanthophylls. One of the carotenes (beta-carotene) is used by some animals in the formation of vitamin A. Therefore yellow and green vegetables and fruits which contain beta-carotene are vital in the diet of some animals.

Remove a small piece of red pepper skin, put it on a clean slide and scrape thin areas using a single edge razor blade or scalpel. Add a drop of water and prepare a wet mount. Observe, draw and label a cell containing chromoplasts.

2. Animal Cells.

a. Cheek cell (squamous epithelium)

The plant cells you have observed have all possessed a cell wall, a cell membrane, cytoplasm, a nucleus, and in some cases, plastids. Now you will have the opportunity to observe some animal cells. Prepare a wet mount of human epithelial tissue by gently scraping the inside of your cheek with the wide portion of a clean toothpick and placing the scrapings in a drop of water on a clean slide. Coverslip the preparation and observe it under low power. Move the slide until you locate cheek cells. Center the cells and view them under high power. Diagram a few cells below and label nucleus, cytoplasm and cell membrane.

Compare the shape of these cells with one another and with the plant cells you previously observed. The epithelial cells have an irregular shape while the plant cells generally exhibit a uniform rectangular shape. Is there any evidence of a cell wall? The cell wall in a plant cell helps in the maintenance of its uniform shape. The epithelial cell, like all animal cells lacks a cell wall.

Stain the specimen with methylene blue dye by placing a drop at the edge of the coverslip. The nucleus should now appear dark blue. Are any plastids present? Since only plant cells contain plastids, they are absent in epithelial cells.

b. Nerve cell or neuron

Examine a prepared slide of the motor neurons from the spinal cord of an ox. Search around for a large nucleated cell with many projections. This is a neuron. Estimate the maximum size of the largest neuron you find by the method of estimating the per cent of the field that the cell occupies (see Exercise 2). Neurons are considered to be large animal cells.

C. A Closer Look at Cells

In almost all of the cells you have examined so far you have been able to see only a small portion of the cellular detail. Since the limit of resolution of the light microscope is about 0.2 μm (= 200 nm), smaller structures cannot be seen. By using an electron microscope (with a limit of resolution of approximately 0.5 nm) the minute details of the cell and its organelles can be examined.

1. Observe the electron micrographs on demonstration and identify the cell parts described below.

a. *cell (plasma)* membrane (7–10 nm in width) Note that the membrane is made up of two layers. Each layer is composed of phospholipids interspersed with a variety of protein molecules. The function of the cell membrane is to regulate the movement of molecules into and out of the cell. Can an unstained cell membrane be resolved with the

light microscope? _____

b. *chloroplast* (2–4 μm in length) This large organelle is surrounded by *two* membranes. A third membrane is arranged in a complex series of parallel layers (lamellae). Certain areas of the membrane system appear as stacks (grana) of disk-shaped membranes. The chloroplast is found only in plants and protists and functions as the site for the process of photosynthesis (the conversion of light energy into chemical energy).

c. *mitochondrion* (2–8 μm in length) This sausage-shaped organelle possesses a double membrane and plays a vital role in the synthesis of ATP. The inner membrane extends inward as a series of fingerlike folds called cristae.

d. *endoplasmic reticulum* This is a series of membrane bound channels running throughout the cell. It functions in internal transport, protein synthesis (rough e.r.) and lipid synthesis (smooth e.r.).

e. *ribosomes* (25 nm in diameter) These particles are composed of ribosomal RNA and protein and are the sites for protein synthesis. Ribosomes occur free in the cytoplasm or attached to the endoplasmic reticulum.

f. *Golgi complex* It is composed of elongated membrane sacs stacked on one another that serve as "processing" areas. Within the Golgi complex proteins are combined with sugars to form glycoproteins. Sugar units are joined together to form a storage product called glycogen, and lipids are combined with proteins or carbohydrates to form lipoproteins and lipopolysaccharides, respectively.

If you carefully observe the peripheral edges of the Golgi complex you can see very small round membranous sacs. These sacs are formed when a portion of the Golgi complex pinches off. The sacs contain groups of processed molecules, some of which will be used

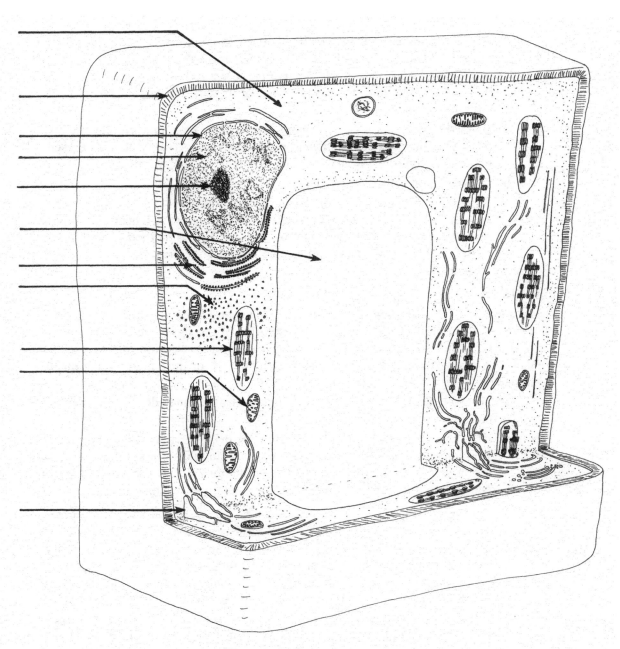

Figure 6-1. Plant cell.

within the cell, while others will be secreted to the outside. Cells, such as liver and intestinal cells, containing an extensive Golgi complex are very active in secretion.

g. *lysosomes* (.5 μm in diameter) Some of the sacs released from the Golgi *complex* contain very powerful hydrolytic (digestive) enzymes specialized for internal (intracellular) action. Such sacs are called lysosomes. When a food vacuole forms in the cell, a lysosome fuses with it and the enzymes contained within the lysosome break down the food molecules. The enzymes are also released when cells are damaged aiding the removal and repair of tissues.

h. *microtubules* These tiny tubes are composed of protein subunits which are linked together. Microtubules function in giving support and form to a cell. During cell division microtubules form *spindle fibers* which aid in the alignment and distribution of the genetic material (chromosomes). Microtubules also are part of specialized structures called *cilia* and *flagella*. These function in the movement of cells and in moving particles

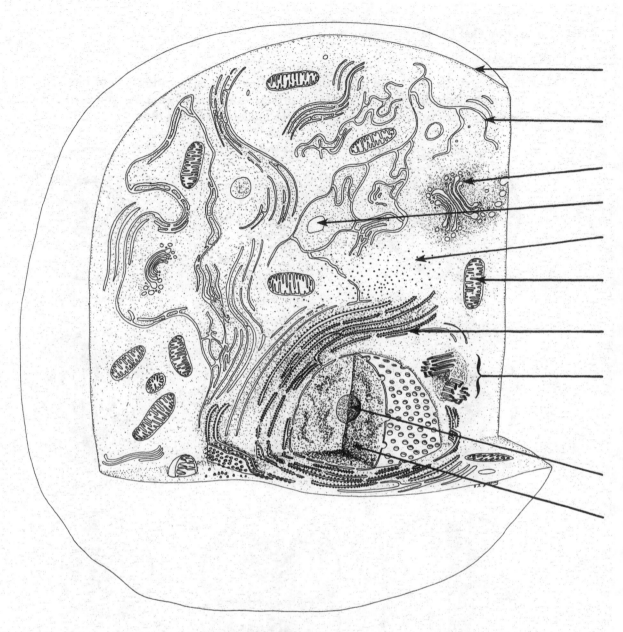

Figure 6-2. Animal cell.

past cells. *Centrioles* and the *basal bodies* of cilia and flagella also contain microtubules. Describe the arrangement of microtubules in:

Cilia and Flagella! _____

Centrioles and basal bodies! _____

i. *nucleus* The nucleus is a large spherical structure bounded by a porous double membrane. Within the nucleus one or more smaller bodies can be seen. Each body is a *nucleolus* (plural: nucleoli). Each nucleolus consists of protein, ribosomal RNA and a segment of DNA active in rRNA synthesis.

The granular material occupying the interior of the nucleus is composed of protein and DNA and is collectively referred to as *chromatin material*. Because it contains DNA, the nucleus is referred to as the control center of the cell.

2. Label the plant and animal cells diagrammed in Figures 6–1 and 6–2.

53

III. Eucaryotic Cell Cycle

A. Phases

The eucaryotic cell cycle is the time period extending from the end of one cell division to the end of the next division. It includes interphase and cell division. As you read the descriptions of the five phases refer to Figure 6-3 and to models depicting them.

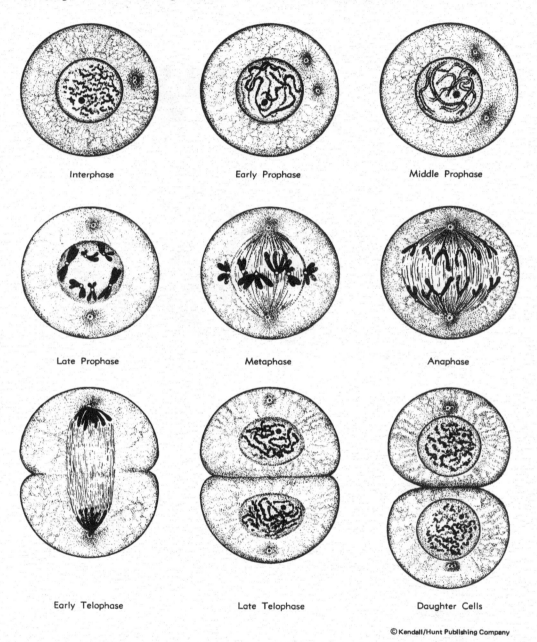

Figure 6-3. Cell division in animal cells.

1. *Interphase* (Inter = between)

   A cell spends most of its time during the cell cycle in interphase. It is in between divisions, and is quite active metabolically. It is growing and replicating its DNA in preparation for the next division. Under the light microscope a distinct nucleus is evident, the nuclear material, chromatin, appears granular, and one or more nucleoli are visible. Paired centrioles are located next to the nucleus in animal cells but are not visible under the light microscope.

2. Cell Division

   Cell division consists of two events: division of the nucleus, *mitosis,* and division of the cytoplasm, *cytokinesis.*

   a. Mitosis

   Mitosis consists of four phases: prophase, metaphase, anaphase, and telophase. During the phases of mitosis identical sets of chromosomes are distributed to two new nuclei.

   (1) *Prophase*

   During prophase the nuclear envelope breaks down allowing the cytoplasm and nuclear material to mix. The nucleoli disperse and disappear from view. Chromatin condenses and coils up to form a characteristic number of rod like *chromosomes* visible under the light microscope. Each chromosome consists of two identical *chromatids* joined at a common *centromere*. In animal cells, paired centrioles migrate to opposite ends of the cell and produce *spindle fibers* and *astral rays*. In most other cells spindle fibers form apparently in the absence of centrioles. The centromere of each double stranded chromosome becomes attached to a spindle fiber from each end of the cell and begins to move toward the center of the spindle.

   (2) *Metaphase*

   The centromeres of all of the double stranded chromosomes are aligned across the center of the spindle.

   (3) *Anaphase*

   Centromeres divide and single stranded chromosomes, formerly called chromatids, move toward opposite ends of the cell.

   (4) *Telophase*

   Telophase begins the instant the chromosomes reach the respective ends. Chromosomes uncoil and spread out as chromatin. The spindle disappears and the nuclear envelope and nucleoli reappear. When telophase ends two interphase nuclei are evident.

   b. Cytokinesis

   The cytoplasm is divided into two approximately equal portions often beginning during anaphase and ending more or less simultaneously with telophase. In animal cells a *cleavage furrow* forms as the cytoplasm is pinched in two by a ring of contracting microfilaments. In higher plants a *cell plate* forms across the center of the cell and proceeds to partition the cytoplasm in two. The cell plate becomes the middle lamella, which is a layer common to the cell walls of the daughter cells. A primary wall is laid down by each cell on its side of the middle lamella.

B. Dynamic Aspects of Cell Division

View the video entitled "Mitosis: Sending the genetic message" to observe the appearance and movement of chromosomes and cytokinesis during cell division.

C. Microscopic Observations of Cell Division

Obtain and examine prepared microscope slides of whitefish blastodiscs (animal cells) and of *Allium* (onion) root tip l.s. (plant cells). Locate the five phases in each cell type and draw and label them in the space provided. Virtually every cell of a whitefish blastodisc should illustrate one of the phases. In the onion root tip look for cells in the area of rapid cell division (meristem) just behind the root cap.

Phase	Whitefish	Onion
Interphase		
Prophase		
Metaphase		
Anaphase		
Telophase		

MEIOSIS AND GENETICS

Objectives

Upon the completion of this exercise you should be able to:

1. Distinguish between the mitotic events in cell division and the events of meiosis.
2. Identify homologous chromosomes and describe the processes of synapsis, crossing over and independent assortment.
3. Illustrate heterozygous and homozygous genotypes and list the kinds of gametes each type produces.
4. Diagram genetic crosses involving traits whose mode of inheritance is simple recessive, complete dominance, incomplete dominance, co-dominance or multiple allelic.
5. Identify a test cross and explain its use.
6. Diagram genetic crosses involving one or more gene locations (monohybrid, dihybrid, etc.).
7. Identify the procedure and theory involved in blood typing and interpret the results of a blood typing test.
8. Determine the mode of inheritance of a trait from pedigree data.
9. Solve all of the review problems in the exercise.

NOTE: YOUR INSTRUCTOR WILL ASSIGN ALL OR PARTS OF THIS LABORATORY EXERCISE FOR YOU TO COMPLETE.

I. Introduction

For many centuries man has known that certain characteristics can be inherited. Application of this knowledge has been used to modify forms to improve vigor, nutritional value, and beauty. Selective breeding of animals, such as beef and dairy cattle, chickens, dogs and cats also has been utilized to meet man's needs and often his whims. Although much practical work has been performed, the fundamental mechanism by which inheritance is governed was not discovered until 1866. At that time an Austrian monk by the name of Gregor Mendel published a paper in which he attempted to describe the inheritance of certain characteristics of the garden pea plant (*Pisum sativum*). Unfortunately, his work was not given much consideration by his fellow scientists. It was not until 1900 that Mendel's work was rediscovered by three botanists in three different countries who were conducting research on plant hybridization. A few years later, the process of meiosis was shown to support Mendel's work. In this laboratory exercise you will be introduced to the process of meiosis and some of Mendel's basic principles. You also will have the opportunity to explore certain aspects of human genetics.

II. Meiosis

A. Meiosis and Sexual Reproduction

Meiosis is a special kind of cell division associated with the process of sexual reproduction. In animals meiosis results in the production of sex cells (eggs and sperm). In higher plants meiosis results in the production of spores which in turn give rise ultimately to the sex cells. A cell undergoing meiosis begins by duplicating its genetic material and then going through two divisions (meiosis I and meiosis II), resulting in the production of four daughter cells, each containing half the amount of genetic material as the original cell. Meiosis also provides a mechanism for maintaining a constant chromosome number for each species and for providing a tremendous amount of variation among offspring. You might imagine that to accomplish all these tasks, meiosis must be a very complex process. Fortunately, it is easily understood if one is familiar with regular cell division. Therefore, let us first review the main events of regular cell division.

B. Cell Division High-lights

In your study of cell division (exercise 6), you discovered that a dividing cell first duplicated its genetic material and then proceeded through several mitotic events in which:

1. The chromatin material was packaged in the form of double stranded chromosomes (i.e., two chromatids);

2. The chromosomes individually lined up along the center of the cell; and

3. The centromeres divided, allowing the single stranded chromosomes (composed of one chromatid) to move to opposite ends of the cell where each uncoiled to form chromatin material again. Finally, a separation of the cytoplasm (cytokinesis) occurred producing two genetically identical daughter cells.

C. Meiosis High-lights

1. Chromosome construction

You may understand meiosis better if you simulate chromosome movement using clay chromosomes of different colors. Obtain enough clay of one color to make two chromosomes, one about seven centimeters in length and the other about ten centimeters in length. Split each chromosome lengthwise to simulate two *sister* (identical) *chromatids*. Pinch the short chromosome in the middle to simulate the centromere which holds the two sister chromatids together. Pinch the long chromosome together at a position approximately three centimeters from one end to simulate its centromere. Next obtain clay of a different color and construct two more chromosomes. Make sure that one of the new chromosomes is identical (except for color) to the short chromosome you already made, and that the other new chromosome is identical to the long chromosome.

Each chromosome consists of two sister chromatids joined together at the centromere. You now should have two sets of double stranded chromosomes; one set consisting of a short and long chromosome of one color; and a second set consisting of a short and long chromosome of a different color. Cells containing two sets of chromosomes are called *diploid*. The body cells of a human are diploid, with each set consisting of 23 different chromosomes. The diploid condition usually is designated as *2n*, where "n" represents the number of chromosomes in each set. In humans, n = _____ . In your clay simulation, n = _____ .

2. Synapsis and homologous chromosomes

The differences between mitosis and meiosis begin during Prophase I (the prophase in meiosis I). The chromosomes of meiotic cells undergo a special pairing-up process called *synapsis*. To illustrate synapsis, take the two short chromosomes and bring them together so that they pair up all along their length. Repeat the process with the two long chromosomes. The two short chromosomes paired up with each other because they are *homologous*, that is, they are the same length, their centromeres are in the same position and they contain the same genes (loci). Therefore, synapsis is the pairing-up of homologous chromosomes. How many homologous pairs of clay chromosomes do you have? _____ . A human cell undergoing meiosis contains how many homologous pairs? _____ .

62

3. Crossing over

Crossing over is the exchange of portions of chromosomes between *non*sister chromatids. Note that each synaptic pair of chromosomes consists of two chromatids of one color and two chromatids of another color. The chromatids of different colors are called nonsister chromatids. To demonstrate crossing over, select the long chromosome pair. Make a cut approximately two centimeters from the long end of two nonsister chromatids and switch (cross over) their positions.

4. Lining up and separating

By the end of Prophase I, the homologous chromosomes have synapsed and some crossing over has occurred. Now the chromosome pairs are ready to line up along the equator (middle) of the cell. Unlike mitosis, where individual chromosomes line up, in meiosis, chromosome *pairs* line up. This stage of meiosis is called Metaphase I. Line up your clay chromosome pairs. Assume that one pole of the cell is to your right and the other pole is to your left. Draw your alignment in the space below:

In Anaphase I the homologous pairs separate and move to opposite poles. The centromeres DO NOT divide at this time. Demonstrate this activity with your clay chromosomes. At each pole you should have one long and one short double stranded chromosome. This condition corresponds to Telophase I. At this time cytokinesis occurs and two daughter cells are formed. Note that each daughter cell has one-half the number of chromosomes as did the original cell. The original cell had a total of four chromosomes and was diploid (2n). Each daughter cell has only two chromosomes, therefore each is a haploid cell (n).

5. Meiosis II

Both of the haploid daughter cells undergo meiosis II, with NO further duplication of the genetic material. The events of meiosis II are similar to the events in mitosis. Using your clay chromosomes, perform meiosis II with each daughter cell. Diagram the four products you obtain in the space below:

Check the daughter cells you have drawn to ensure that each daughter cell has a long chromosome (consisting of a centromere and one chromatid) and a short chromosome (consisting of a centromere and one chromatid). Note that the four cells are different due to the crossing over in the large chromosome which occurred in Prophase I.

63

6. Independent assortment of chromosomes

Cross overs represent one way in which meiosis contributes to genetic variation. Another way involves how the chromosome pairs line up at Metaphase I and subsequently, how they move apart in Anaphase I. Re-assemble your chromosome pairs as they appeared in Metaphase I. Consult your drawing if necessary. Change the position of one of the pairs so that the chromosomes that make up the pair are NOT facing the same poles as they were before. Take both chromosome pairs through the rest of meiosis and draw the four daughter cells in the space below:

These daughter cells should contain different chromosome combinations than the four previous daughter cells. If they do not, try the procedure again.

In your simulation involving only two pairs of chromosomes, there were two different ways the pairs could line up and move to opposite poles. The movement of the long pair did not influence the movement of the short pair and vice versa. Therefore, each chromosome pair assorts (separates) independently. The manner in which the chromosome pairs line up is simply a matter of chance.

7. Summary

Meiosis begins with a diploid (2n) cell which duplicates its genetic material one time, goes through meiosis I (prophase I, metaphase I, anaphase I and telophase I) and undergoes a cytokinesis producing two haploid (n) cells. Each haploid daughter cell undergoes meiosis II (prophase II, metaphase II, anaphase II and telophase II) and a second cytokinesis, resulting in the production of four haploid cells.

One major event which occurs in meiosis is the pairing-up of homologous chromosomes.

This event is called _____ and occurs in _____ . Two other major events occur that contribute to genetic variation. The first event involves the

exchange of segments between nonsister chromatids and is called _____
_____ . The second event involves the individual lining up and movement of the

chromosome pairs to their respective poles and is called _____ _____ .

Keep the clay chromosomes handy when you explore the next section dealing with Mendelian principles. The genes with which you will be working, are located on chromosomes and you may be able to see how the genes segregate and undergo independent assortment by actually placing gene markers on the clay chromosomes and taking the chromosomes through the process of meiosis.

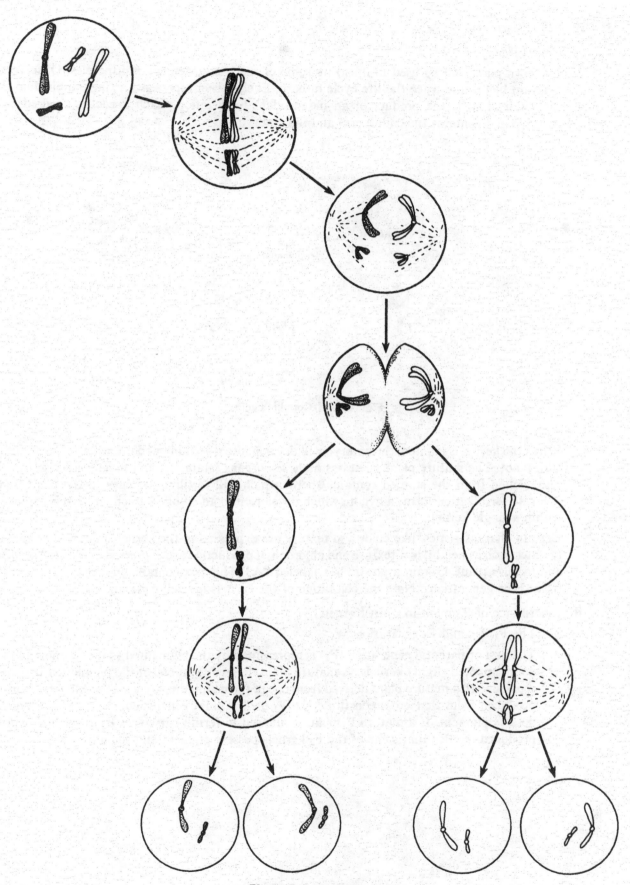

Figure 7-1. Meiosis.

III. Mendelian Genetics

A. Corn Life Cycle

The corn plant will be used to illustrate some of the basic Mendelian principles. Observe the diagram below and trace the life cycle of corn. Each kernel represents a single individual and each individual is the product of an independent fertilization. Each kernel is covered by a pericarp and contains the endosperm and the embryo.

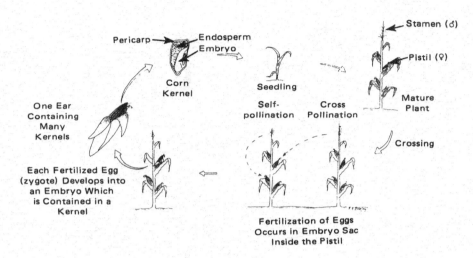

Figure 7-2. Corn life cycle.

When the kernel is planted, the embryo will develop into a seedling and eventually into a mature corn plant. The mature plant possesses male sex organs (stamens) and female sex organs (pistils). Pollen from the stamen contains sperm which will fertilize the eggs in the pistil. Each separate fertilization forms a zygote which develops into an embryo. Each embryo is contained within a single kernel.

Certain characteristics, like color and texture are expressed in the kernel stage. Other characteristics such as height, leaf texture and chlorophyll production are expressed only in the seedling and mature plant. Obtain a plastic box labeled "A." This box contains models of ears of corn from three generations. Note the coloration of the kernels in each generation.

B. Inheritance of Pigment in Corn Kernels

1. The P_1, F_1, and F_2 generations.

   The first or parental generation (P_1) is represented by the pigmented (dark) kernels and the nonpigmented (light) kernels. A mature plant derived from one of the pigmented kernels is crossed with a mature plant derived from one of the nonpigmented kernels. All of the kernels produced from that cross constitute the F_1 or first filial generation. All of the F_1 kernels possess pigment. If a mature F_1 plant is allowed to fertilize itself an F_2 generation will be produced. Notice that some of the F_2 kernels possess pigment and some do not.

2. Alleles, genotypes, and phenotypes.

The pattern of inheritance of pigmentation in corn kernels may be explained by Mendelian theory. According to the theory, the characteristic of kernel pigmentation is controlled by a single gene. Since corn reproduces sexually, each fertilized egg (zygote) must contain two representatives of the gene. (One donated by the male parent and one donated by the female parent.)

The P_1 pigment kernels possess a form of the gene which allows for pigment production. This form of the gene may be represented by the letter "D." Each cell in the P_1 plant produced from a pigmented kernel contains two representatives of the gene (i.e., DD).

The P_1 nonpigmented kernels contain an alternate form of the gene which does not produce pigment. This alternate form may be represented by the letter "d." Each cell in the P_1 plant produced from a nonpigmented kernel contains two representatives of the gene (i.e., dd).

Each form of the gene (D and d) is referred to as an *allele*. The combination of alleles in an organism is called a *genotype*. The outward expression of the genes (or the observable characteristic) is called the *phenotype*. The phenotype of the DD kernel is the presence of

pigment. What is the phenotype of the dd kernel? _____

3. Segregation of alleles.

The mature P_1 plants produce sex cells (gametes) by the process of meiosis. According to Mendelian theory, alleles *segregate* (separate from each other) into different gametes. This means that in each gamete there will be only *one* representative of each gene pair. Therefore the mature DD plant will produce gametes which contain only one "D" allele. In the chart below indicate the type of gametes that will be produced by the mature dd plant.

Phenotype	Pigmented Kernels	Nonpigmented Kernels
genotype	DD	dd
	meiosis	meiosis
gametes	Ⓓ	◯

4. Dominant and recessive alleles.

When the pigmented plant is crossed with the nonpigmented plant, the male gametes from one plant combine with the female gametes of the other plant to form the fertilized egg or zygote. For example, if the female gamete from the pigmented plant Ⓓ were fertilized by a male gamete from the nonpigmented plant ⓓ, the F_1 plant would have the genotype

Dd . What is the phenotype of the kernels of the F_1 generation? _____

What is their genotype? _____ The kernels of the F_1 generation are all pigmented. Since the "D" allele is expressed, it is referred to as the *dominant* allele. The "d" allele is not expressed, therefore it is referred to as a *recessive* allele.

5. Heterozygous and homozygous genotypes.

Notice that the genotype of the F_1 plants contains a "D" allele and a "d" allele. When a genotype possesses two different alleles (Dd) the genotype is termed *heterozygous*. When the genotype possesses two identical alleles (DD or dd) it is termed *homozygous*.

6. The F_1 cross.

Two heterozygous F_1 plants may be crossed to produce the F_2 generation. In this situation the F_1 plants serve as the parents for the F_2 generation. The cross may be diagrammed as indicated below:

	Female Parent	Male Parent
(1) List the *phenotypes* of each F_1 plant which will serve as a parent for the F_2 generation.	pigmented kernels	pigmented kernels
(2) List the *genotypes* of each parent.	Dd	Dd
(3) Determine the different types of *gametes* that may be produced by each parent. Since each heterozygous genotype contains two *different* alleles, two different types of gametes may be produced.	Meiosis ⒟ ⒟	Meiosis ⒟ ⒟

(4) Determine *all* the possible ways in which the female gametes may combine with the male gametes. One method of accomplishing this is through the use of a *Punnett Square*. To construct a Punnett Square, list all of the female gametes in a vertical column. Then list all of the male gametes in a horizontal row to the right and just above the female column.

Female Gametes	Male Gametes	⒟	⒟
⒟			
⒟			

To complete the square, just fill in the empty spaces. For example, if the female gamete ⒟ were fertilized by the male gamete, ⒟ , the resulting zygote would have the genotype DD. Complete the Punnett Square above by indicating the genotypes and phenotypes resulting from each possible gametic union.

(5) List the F_2 progeny giving genotypes, phenotypes, and the expected frequency of each type.

Expected F_2 genotypic frequencies	25% DD	50% Dd	25% dd
Expected F_2 phenotypic frequencies	75% pigmented kernels		25% non-pigmented kernels

Note: These frequencies also may be expressed as:

1/4 DD	2/4 Dd	1/4 dd		1DD	2Dd	1dd
¾ pigmented kernel		¼ non-pigmented kernel	OR	3 pigmented kernels		1 non-pigmented kernel

Count the number of pigmented kernels and the number of non-pigmented kernels on the F_2 ear of corn in plastic box "A." The kernels on this ear represent the F_2 generation. Calculate the frequency of each phenotype by dividing the counted number in each group by the total and multiplying the result by 100.

Group	Number
Pigmented	
Nonpigmented	
Total	

Group	$\dfrac{\text{Number}}{\text{Total}} \times 100$	%
Pigmented		
Nonpigmented		
Total percentage		100.0

Compare the observed F_2 frequencies with the expected F_2 frequencies.

7. The test cross.

Predict the expected genotypic and phenotypic ratios when an F_1 heterozygous pigmented kernel plant (Dd) is crossed with a homozygous nonpigmented plant (dd). Diagram the cross:

Phenotypes	Female Pigmented Kernels	Male Nonpigmented Kernels
Genotype		
	Meiosis	Meiosis
Gametes	◯ ◯	◯

F₂ genotypic ratio _____

F₂ phenotypic ratio _____

Check your prediction by counting the F_2 kernels in plastic box "B." When a plant exhibiting a dominant characteristic (e.g., pigmented kernels) is crossed to a *homozygous recessive* plant (nonpigmented—dd), the cross is called a *test cross*. If the plant expressing the dominant characteristic is heterozygous, the test cross offspring will exhibit both the dominant and recessive phenotype in a 1:1 ratio.

C. Inheritance of Leaf Color in Tomato

Observe the leaf color of the F_2 tomato seedlings on demonstration. How many phenotypes are present? _____ What is the phenotypic ratio among the seedlings? green seedlings _____ : greenish-yellow seedlings _____ : yellow seedlings _____ Using the genotype Gg for both F_1 individuals, diagram the cross which will produce the F_2 generation and compare the expected F_2 phenotypic ratio with the F_2 phenotypic ratio which you actually observed from the tomato seedlings.

F_1 genotypes	Female Gg		Male Gg	
Genetic makeup of gametes	◯	◯	◯	◯
Expected F_2 genotypic frequencies				
Expected F_2 phenotypic frequencies				

When two alleles in the heterozygous condition produce a phenotype which is *intermediate* between the phenotype of the two homozygous genotypes, the mode of inheritance is called *intermediate inheritance* or *incomplete dominance*.

D. The Dihybrid Cross and Independent Assortment

A cross involving only *one* set of contrasting traits (e.g., pigmented vs. nonpigmented kernels) is called a *monohybrid cross*. A cross involving *two* sets of contrasting traits is called a *dihybrid cross*.

Obtain an ear of corn labeled "F_2." Four different phenotypes are present on the ear. In the space below, describe each phenotype in reference to color and texture:

            Color   Texture

1. _____   _____

2. _____   _____

3. _____   _____

4. _____   _____

Two different traits are represented on the ear of corn: *color* and *texture*. Each trait is controlled by a separate gene. The gene that controls color is located on a different chromosome than the gene that controls texture. For example, using the clay chromosomes, designate the location (locus) of the color gene to be on the long chromosome and the texture gene to be on the short chromosome.

Use the letter "R" to represent the color gene and the letter "S" to represent the texture gene. The dark (purple) kernels are produced by the dominant allele, "R." The light (yellow) kernels are due to the presence of the recessive allele, "r." Therefore all yellow kernels must be homozygous "rr." At the "S" gene locus, the dominant allele, "S" produces a smooth texture (starchy) and the recessive allele, "s" produces a wrinkled texture (sweet). What is the genotype of a yellow wrinkled kernel? _____ . Since two separate genes are involved, both must be acknowledged in the kernel's genotype: "rr,ss." What is the genotype of a kernel homozygous for purple and smooth? _____ .

In the space below, diagram a cross through the F_1 generation, beginning with a true breeding (homozygous) plant (female) that was produced from a purple, smooth kernel (RR,SS) and a plant (male) produced from a yellow, wrinkled kernel (rr,ss). NOTE: Each gamete must contain one representative (allele) from each gene locus: (R,S) from one parent and (r,s) from the other parent.

Parental Phenotypes (List both kernel color *and* endosperm texture for each parent)	Female	Male
Parental genotypes		
Genetic make-up of parental gametes	◯	◯

Punnett Square

Female Gametes \ Male Gametes	◯
◯	F_1 genotype F_1 phenotype

All of the F_1 offspring have the genotype Rr,Ss with a purple, smooth phenotype. Next, in the space below, diagram a cross between two F_1 plants to produce an F_2 generation. HINT: Place the gene symbols on the clay chromosomes and move them through meiosis in order to determine the gametes produced by each F_1 plant. Remember the chromosome pairs line up independently of one another. There are two possible ways in which the two chromosome pairs can line up. Both ways must be considered in gamete production. This results in the possibility of four different types of gametes being produced by each F_1 plant. Each gamete will contain one allele of each gene, that is one *"R"* (R or r) *and* one *"S"* allele (S or s). The term, *independent assortment,* is used to identify the independent movement of chromosomes bearing genes into different gametes.

	F_1 Male	F_1 Female
Phenotype	_____	_____
Genotype	_____	_____
Gametes	◯ ◯ ◯ ◯	◯ ◯ ◯ ◯

The F₁ male flower parts produce the following gametes:

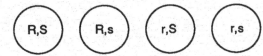

The F₁ female flower parts produce the same types of gametes because the genotype is identical to the F₁ male parts. Complete the Punnett Square below to determine the expected genotypic and phenotypic frequencies of the F₂ generation.

Punnett Square

Female Gametes	Male Gametes			
	◯	◯	◯	◯
◯				
◯				
◯				
◯				

Expected F₂ Genotypic Frequencies: Transfer and organize the data from the Punnett Square to the chart below.

Genotype	Frequency		Genotype	Frequency		Genotype	Frequency

Nine different genotypes are possible.

Expected F₂ Phenotypic Frequencies: Some of the genotypes in the chart above produce the same phenotype. For example, all genotypes which contain at least one dominant allele at each locus (RR,SS; RR,Ss; Rs,SS; Rr,Ss) result in a purple, smooth phenotype. The genotypes may be collectively expressed as R_,S_.

List the four F_2 phenotypes and the expected frequency of each, in the chart below:

Genotype	Phenotype (color, texture)	Frequency (proportion)	%
R_,S_	_____	_____	56.25
R_,ss	_____	_____	18.75
rr,S_	_____	_____	18.75
rr,ss	_____	_____	6.25
		TOTAL: 16 / 16	100.00

Observed F₂ Phenotype Frequencies:

Next, count the kernels representing each phenotype on the ear of corn labeled "F_2" and determine the phenotypic frequency. Record the data in the chart below:

Phenotype	Total	Frequency (Total / GRAND TOTAL)	% (Freq. × 100)
purple, smooth	_____	_____	_____
purple, wrinkled	_____	_____	_____
yellow, smooth	_____	_____	_____
yellow, wrinkled	_____	_____	_____
GRAND TOTAL:	_____	1.00	100.00

Do the observed percentages agree with the expected percentages? Chi Square Test information may be available at the demonstration desk that can be used to statistically test the "goodness of fit" of your results.

IV. Human Genetics

A. Human Pedigrees.

1. Pedigree symbols.

In determining the method of inheritance of a particular trait in humans, genetic counselors will sometimes use pedigree analysis. Each individual in the pedigree is represented by a symbol. Family relationships are indicated by connecting lines. Below are some of the symbols used in the construction of human pedigrees.

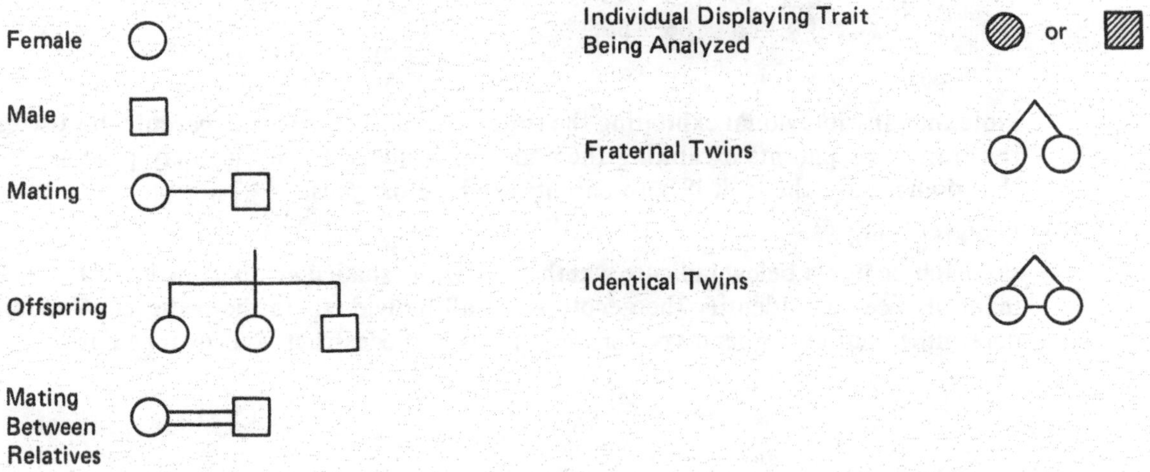

2. A pedigree of a recessive trait.

The following pedigree illustrates the inheritance of the recessive trait, albinism (lack of pigment production).

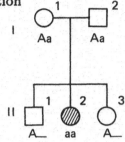 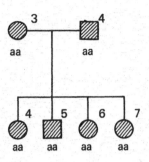

Generation I

Generation II

Genotypes	Phenotypes
AA	normally pigmented
Aa	normally pigmented
aa	albino

Note that individual II-2 is an albino (aa), but both of her parents (I-1 and I-2) are normally pigmented. When both parents do not express the trait, but produce an offspring that does express the trait, the trait is *recessive*. If the trait is recessive, two parents that possess the trait (I-3 and I-4) will produce only offspring that have the trait. Note that the symbol A__ is used when it cannot be determined from the information given whether an individual is AA or Aa.

3. A pedigree of a dominant trait.

The following pedigree illustrates the inheritance of the dominant trait, polydactyly (extra fingers and/or toes).

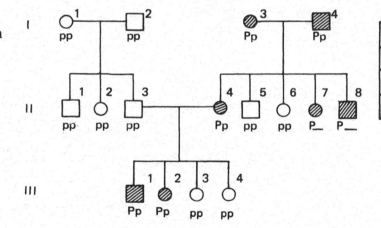

Generation I

Generation II

Generation III

Genotypes	Phenotypes
PP	Polydactyly
Pp	Polydactyly
pp	Normal

Note that any individual exhibiting the trait must have at least one parent who also exhibits the trait. Two parents exhibiting the dominant trait (I-3 and I-4) may produce offspring that do not have the trait if both parents are heterozygous.

4. Pedigree analysis.

For each pedigree below indicate whether the trait (shaded symbols) is inherited as a dominant or recessive. Identify the genotypes of all individuals in the pedigree. You may also determine your own phenotype for these traits and construct your own pedigree.

Pedigree 1. Tongue Rolling (ability to roll tongue into a U-shape)

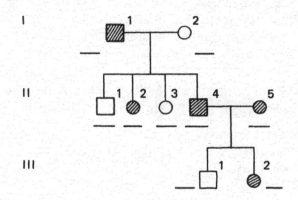

mode of
inheritance: _____

Pedigree 2. Dry ear wax (ear wax is dry and crumbly instead of wet and sticky)

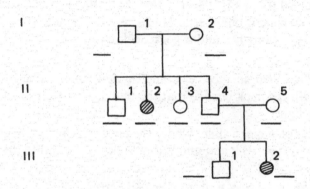

mode of
inheritance: _____

B. Human Blood Groups.

  1. The ABO system.

In the ABO system, individuals are classified on the basis of the antigens (antibody stim-ulating factors) located on the membranes of their red blood cells. Type A individuals pos-sess the A antigen; type B individuals possess the B antigen; type AB individuals possess both A and B antigens; and type O individuals possess neither A nor B antigens. Blood typing is important in transfusions because of naturally occurring antibodies in the blood serum. These antibodies will react with specific antigens on the red blood cell membrane and cause these cells to clump together (agglutinate). For example, individuals with type A blood (have the A antigen) possess antibodies (anti-B) in their serum which will agglu-tinate cells which possess the B antigen. Using this principle match the following blood types with their antigens and antibodies:

Blood Type	Antigen on Red Blood Cell	Antibody in Serum
A	antigen A	anti-B
B		
AB		
O		

Answer Choices
antigen A
antigen B
antigens A and B
neither A antigen nor B antigen
----
anti-A
anti-B
anti-A and anti-B
neither anti-A nor anti-B

2. The Rh System.

The Rh system is a completely separate blood group system from the ABO system. Individuals who are Rh positive possess the Rh antigen on their red blood cells. Individuals who are Rh negative do not possess the antigen. No natural antibodies are present in the serum, but once an Rh negative individual is exposed to Rh positive cells, the individual will be able to produce anti-Rh antibodies.

Production of the Rh factor is controlled by the dominant allele "Rh". What is the probability that an Rh$^-$ mother (rh,rh) and a heterozygous Rh$^+$ father (Rh,rh) will produce an

Rh$^+$ child? _____ What are the dangers to an Rh$^+$ fetus if the mother previously has been sensitized to produce anti-Rh antibodies? Note that antibodies are able to cross the placental barrier and enter the fetal circulation.

3. Blood typing procedure.

At this time you may wish to type your own blood for the ABO system and the Rh factor. In the next section you will study the inheritance of the blood groups. The following materials will be needed for the typing procedure: a microscope, two clean slides, five clean toothpicks, a 0.9% saline solution, a wax pencil, one sterile lancet, ethyl alcohol, cotton, anti-A serum, anti-B serum, anti-Rh serum (anti-D), and a gooseneck lamp.

Technique:

   a. The Rh determination will require a warm surface. Use an Rh-view box or place a gooseneck lamp on a dark surface with the neck bent so that the bulb is about seven cm above the surface and switch on the light.

   b. Using a wax pencil, draw three circles on a clean slide. Under the circle on the left label "anti-A"; under the middle circle label "anti-B"; and under the circle on the right label "control."

   c. Place one drop of anti-A serum in the left circle, a drop of anti-B serum in the middle circle, and a drop of 0.9% saline solution in the circle on the right.

   d. On a second clean slide draw two circles and label the left hand circle "anti Rh" and the right hand circle "control." Place a drop of anti-Rh serum in the left hand circle and a drop of 0.9% saline solution in the control circle.

   e. Dip a small piece of cotton in the ethyl alcohol and swab the tip of your ring finger. Allow the finger to air dry. Then, using the sterile lancet, prick the tip of the finger and allow the first few drops of blood to run off. Discard the lancet.

   f. Using a clean toothpick transfer the next drop of blood to the drop of anti-A serum and mix the two with the toothpick. Using another clean toothpick, transfer a drop of blood to the anti-B serum and mix. Repeat the process for the control solution. Add a drop of blood to the anti-Rh serum and a drop to the control solution. Use a clean toothpick each time. Place the Rh slide on the Rh-view box or under the light for about four minutes. *ABO Determination*—Compare both mixtures (anti-A and anti-B) with the control and observe for agglutination (clumping) of cells. If the cells in the anti-A serum agglutinate, they must possess the A antigens on their membranes. If the cells in the anti-B serum agglutinate, they must possess the B antigen. If both antigens are present, agglutination will occur in both serums. Conversely, if both antigens are absent, neither side will agglutinate. Indicate your ABO type: _____ , and record it on the chart at the instructor's desk. *Rh Determination*—After the Rh slide has been warmed for about four minutes, observe it under the microscope using low power. Your cells possess the Rh factor if you observe many clumps of at least five cells. Indicate whether you are Rh negative or Rh positive: _____ . Record your blood type on the chart at the instructor's desk.

g. Clean up and place all blood contaminated objects in the disposal container provided.

4. Inheritance of the ABO blood group.

The ABO system is controlled by a single gene pair. It differs from the gene pairs you have previously studied in that the ABO blood system consists of more than two alternate forms (alleles) of the gene. This system, referred to as a *multiple allelic system,* consists of three major alleles and several minor alleles. The symbols traditionally used for the major three alleles are: I^A, I^B, and i.

Even though the ABO system consists of multiple alleles, an individual possesses only two alleles, one of which is inherited from the father and the other from the mother. Since the three major alleles exist in combinations of two, there are six possible genotypes. List all six possible genotypes: _____ , _____ , _____ , _____ , _____ , _____ . For each genotype indicate the blood type (A, B, AB, or O) it will produce.

Genotype	Phenotype (Blood Type)	Genotype	Phenotype (Blood Type)
$I^A I^A$		$I^B I^B$	
$I^A i$		$I^B i$	
$I^A I^B$		ii	

The I^A allele produces the A antigen, the I^B allele produces the B antigen, and the i allele produces neither one. Therefore individuals with "A" type blood may have the genotype $I^A I^A$ or $I^A i$. Individuals with blood type "AB" must have the genotype $I^A I^B$ and "O" individuals must be ii. Note that the allele i is recessive to both I^A and I^B. When the alleles I^A and I^B are together in the genotype $I^A I^B$, both are expressed equally producing the AB blood type. When two alleles occur together and each is expressed equally they are said to be *codominant.*

Predict the possible blood types of the children produced from the following marriages and give the probability for each blood type:

Parents	Possible Blood Types of Children and Probabilities of Each
$I^A I^B \times I^A I^B$	
$I^A i \times ii$	
$I^A I^B \times ii$	
$I^A i \times I^B i$	

GENETICS REVIEW QUESTIONS

1. The outward expression of a gene is called the _____ .

2. Alternate forms of the same gene are called _____ .

3. What gametes may be produced by the following individuals:

   a. AA _____   c. AABb _____

   b. Aa _____   d. AaBb _____

4. Set up a Punnett Square to determine all possible combinations of gametes produced by the following parents: Aa × Aa

5. Diagram through the F_2 generation a cross between a plant that produces smooth seeds (SS) and a plant that produces wrinkled seeds (ss).

6. Diagram an example of a test cross.

7. A cross between two individuals that differ by *two* contrasting traits is called a _____ (monohybrid, dihybrid) cross.

8. Write the pedigree symbols for the following:

   a. a man

   b. fraternal twins

9. Identify the genotypes of all of the members of the following pedigree. Use the symbols AA, Aa, A__ and aa.

10. John has blood type AB. He may receive a transfusion of what blood types? Explain.

11. A man with the blood type A marries a woman with blood type B. Their first child has blood type O. What other blood types may be expressed in their children?

SUPPLEMENTAL PRACTICE PROBLEMS

Monohybrid Cross

1. The Austrian monk, Gregor Mendel crossed a true breeding (homozygous) pea plant with tall stems to a true breeding plant with short stems. All of the other plants in the F_1 (first generation) were tall. From this information, determine which characteristic (tall stems or short stems) was the dominant trait. Cross two F_1 plants and predict the genotypic and phenotypic ratios of the F_2 generation. Diagram the cross.

2. In the garden pea plant, the color of the seed pod is determined by the "G" gene locus. The dominant allele, G, produces green pods and the recessive allele produces yellow pods. What are the phenotypes associated with the following genotypes: GG _green_ ,

   Gg _green_ gg _yellow_ .

   Cross a true breeding plant with green pods to a true breeding plant with yellow pods. Diagram the complete cross through the F_2 generation.

3. A heterozygous plant with green pods (Gg) is crossed to a plant with yellow pods (gg). Diagram this cross through one generation.

4. Mendel crossed true breeding round seeded plants with true breeding wrinkled seeded plants. Among the 7,324 F_2 generation plants obtained, 5,474 (74.4%) were round seeded and 1,850 (25.3%) were wrinkled seeded.

   a. Which characteristic is dominant?
   b. What were the genotypes and phenotypes of the F_1 plants that were used to produce the F_2 generation?
   c. Diagram the entire cross.

5. Mendel crossed a round seeded F_2 plant with an F_2 wrinkled seeded plant and observed the following results:

   > 50% of the plants produced had round seeds and
   > 50% of the plants produced had wrinkled seeds.

   What were the genotypes of the F_2 round seeded plant and the F_2 wrinkled seeded plant used in this cross?

6. Mendel crossed *another* round seeded F_2 plant with an F_2 wrinkled seeded plant, but this time all of the plants produced had round seeds. What were the genotypes of the F_2 round seeded plant and the F_2 wrinkled seeded plant?

7. Purple flowers (P) are dominant to white flowers (p). Diagram a cross (*through* the F_2 generation) between a true breeding purple flowered plant and a true breeding white flowered plant.

8. In mink, *brown fur* is controlled by the dominant gene "B". Its recessive allele "b" produces *silver-blue* mink. A brown mink is given to you as a present. You do not know whether it is homozygous (BB) or heterozygous (Bb).

   a. explain how you would "test cross" this individual.
   b. diagram a cross to show what type of offspring you would expect if the brown mink were homozygous.
   c. diagram a cross to show what type of offspring you would expect if the brown mink were heterozygous.

9. In sheep, white is produced by a dominant gene (W). Its allele (w) produces a black individual. A white ram (male) was test crossed to several ewes (females). The total offspring were 3 white lambs and 2 black lambs.

    a. what were the phenotypes of the ewes used in the test cross?

    b. why was this phenotype used?

    c. based on the phenotypes of the lambs, what is the genotype of the original white ram?

10. When plants which produce round-shaped radishes are crossed with each other, all of the F_1 and F_2 plants produce only round-shaped radishes. When plants that produce long-shaped radishes are crossed with one another—all of the F_1 and F_2 plants produce only long-shaped radishes. However, when round-shaped radish plants are crossed with long-shaped radish plants all of the F_1 are oval in shape. a. If radish shape is controlled by one gene locus, what types of plants would you expect from a cross of two oval-radish plants? b. What is the name given to this type of inheritance?

Dihybrid Crosses

1. A *tall* plant which produces *wrinkled seeds* (Tt,ss) is crossed to a *dwarf* plant which produces *smooth seeds* (ttSs). On a separate sheet of paper diagram the cross to show:

    parental genotypes
    parental phenotypes
    parental gametes
    Punnett square
    F_1 genotypes and proportion of each
    F_1 phenotypes and proportion of each

2. Two pea plants are crossed and produce the following F_1 plants: 50% TtSs 50% Ttss

One of the parental plants was tall and produced smooth seeds

    a. what is its genotype? _____

    b. what was the genotype of the other parental plant?

    _____

    c. what was its phenotype? _____

3. In turkeys, the normal *bronze* color is produced by the dominant gene "B". Its recessive allele "b" produces *red* color. An independently assorting dominant gene, "N" produces *normal* feathers. Its recessive allele "n" produces *hairy* feathers.

    a. diagram a cross through the F_2 generation between a homozygous *bronze* turkey with *normal* feathers (BBNN) and a *red* turkey with *hairy* feathers (bbnn)

    b. diagram a cross through the F_1 generation between a *heterozygous bronze* turkey with *normal feathers* and a *red* turkey with *hairy* feathers.

Human Genetics

The following problems involve some recognizable human traits. After working the problems you may wish to determine your own phenotype and genotype. Construct a pedigree chart for each problem.

1. P.T.C. Taster: To some people, the chemical, phenylthiocarbamide, has a strong bitter taste. However there are other individuals who are unable to perceive such a taste sensation.

Tom is able to taste PTC, but his sister, Terri cannot. Both of their parents can taste PTC. What is the most probable mode of inheritance?

Tom is able to taste PTC, but his sister, Terri cannot. Both of their parents can taste PTC. What is the most probable mode of inheritance?

2. Dimples: Kirk Douglas has a prominent indentation (dimple) in his chin, and so does his son Michael. Dimples are inherited as a dominant trait. If Michael's mother does not have dimples, and Michael's wife does not have dimples, what is the probability that Michael's son will *have* dimples?

3. On the appropriate figure below, place your hand (palm down) with the distal thumb joint (joint nearest nail) on the designated point on line "a". Without moving your hand try to hyperextend your thumb to line "b". Those individuals who can hyperextend their thumbs have the trait called *"hitchhiker's thumb."*

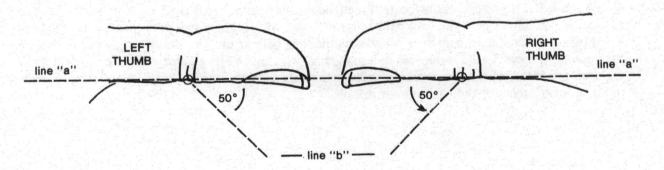

Although the trait basically is controlled by one gene pair, the expression of the trait can vary due to the modifying effect of other genes. This variation often may be observed between the right hand and left hand.

Henry is able to hyperextend his thumb past the 50° mark, but neither parent can do so. What is the most probable inheritance of the hitchhiker's thumb?

4. Attached earlobes: Alice's earlobes are attached directly to the side of her head, but her sister, Freida, has earlobes which hang free below the attachment point. Alice's husband, Allen, also has attached earlobes, as does everyone in his family. Their eight children all have attached earlobes. Freida's husband, Fred, has free hanging earlobes. Fred's mother has free hanging earlobes, but his father has attached earlobes. Of the eight children of Freida and Fred, six have free hanging earlobes and two have attached earlobes. What is the most probable method of inheritance of attached earlobes?

5. Examine the back side (dorsal surface) of your fingers and locate the mid-digital area between the two central joints. Look very carefully and discover if you possess a few hairs in this mid-digital region on any of your fingers.

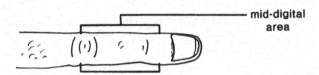

mid-digital
area

Henry possesses mid-digital hairs, but his wife does not. Of their six children, three have mid-digital hairs and three do not. One of Henry's daughters with mid-digital hair grows up and marries a man who also has mid-digital hairs. This couple has two children, none of whom have mid-digital hair. What is the most likely mode of inheritance of mid-digital hair?

6. Long palmar muscle: Leslie has a muscle in her forearm which is called the long palmar muscle. She knows that she has this muscle because she can feel its tendon in her wrist. (You may check to see if you possess this muscle by clenching your fist very tightly and flexing it toward you. Feel the tendons within the wrist. If you can easily locate three tendons, the center one is the tendon of the long palmar muscle. If you do not have the long palmar muscle, you will feel only two prominent tendons). Leslie has a twin brother, Louis, who does not possess the long palmar muscle. Leslie's parents also lack the long palmar muscle. Leslie is married to a man, Jim, who has the long palmar muscle. Their first two children are identical twin boys who also have the long palmar muscle. What is the most probable mode of inheritance for the long palmar muscle?

INTRODUCTION TO TAXONOMY

Objectives

Upon completion of this exercise, you should be able to:

1. Define taxonomy and classification.
2. List in order the seven major taxonomic groups (excluding domains) used in the taxonomic hierarchy.
3. Define species.
4. Describe the binomial system of naming a species.
5. Use a dichotomous key.
6. Construct a dichotomous key.
7. Compare the two, three, four and five kingdom systems of classification with the classification scheme in your textbook.

I. Introduction

There are a great number of different kinds of organisms on the earth. In fact, there are over 1.5 million different species already described. Taxonomy or systematics is a subdivision of biology which deals with describing, classifying, naming and identifying organisms.

A. *Describing Organisms*

Individual organisms are studied and their characteristics listed. Individuals with many characteristics in common are placed in the same species. The species concept is abstract. Species that have common characteristics are considered to be related and are placed in the same genus. The genus concept is abstract and subjective. In a similar manner families, orders, classes, phyla and kingdoms are described. Thus, a taxonomic hierarchy (Fig. 8-1) can be built.

In this hierarchy, the lower the taxonomic group, the more characteristics the organisms in it have in common. Conversely, the higher the taxonomic category, the fewer characteristics all the organisms in it have in common. In this hierarchy, the different species in a genus are related, the different genera in a family are related, etc.

Write your own description of the species we call man or *Homo sapiens*. Remember, only characteristics that individual humans have in common should be included in the description.

B. *Classifying Organisms*

Individuals that are evolutionarily closely related, are structurally, functionally, developmentally and behaviorally similar and which in nature interbreed with one another and not with other kinds of organisms are grouped in the same species. Therefore, a species is a population of interbreeding organisms that is reproductively isolated from other populations of organisms. Closely related species are grouped in the same genus; closely related genera are grouped in the same family and so on. The placing of organisms into categories (taxa) is termed classification. A good, modern classification system enables us to store and retrieve information about organisms in an orderly way. Further, it attempts to place closely related organisms in the same category and to show evolutionary relationships among organisms that are due to a common ancestry.

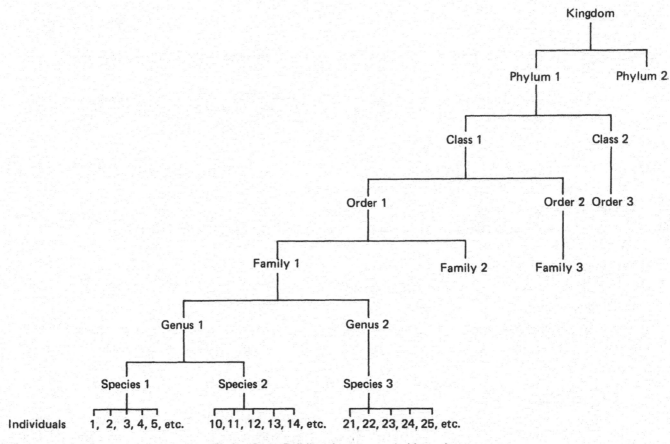

Figure 8-1. Building the taxonomic hierarchy.

It should now be apparent why disagreements arise among taxonomists concerning classification systems. A taxonomist organizes the various groups according to the characteristics that he considers to be most important. The classification systems can and do change as we learn more about each kind of living organism.

Some systems recognize two, three, four or five kingdoms. The oldest systems, dating back to the time of Aristotle, recognized two kingdoms, *Plantae* and *Animalia*. Those organisms that are green or non-motile were considered plants. Animals were non-green but motile. Under this system the sponges and sea anemones were erroneously considered plants! With the collection and assimilation of more data, particularly about cellular details, other systems are now more widely accepted.

The three kingdom system recognizes *Monera* (prokaryotic bacteria), *Plantae* (eukaryotic with plastids and cell walls; includes mosses, liverworts, ferns, gymnosperms and angiosperms), and *Animalia* (eukaryotes lacking plastids and cell walls; includes sponges, jellyfish, flatworms, roundworms, segmented worms, mollusks, arthropods, echinoderms and chordates). Some four kingdom systems include *Monera, Protista* (primarily eukaryotic unicellular forms such as protozoans and algae), *Plantae* and *Animalia*. Others recognize *Monera, Fungi* (eukaryotes lacking chlorophyll and whose cell wall is not composed of cellulose, such as slime molds, molds, yeasts and mushrooms), *Plantae* and *Animalia*. The major difference between the four and five kingdom systems is the recognition of both *Fungi* and *Protista* as separate kingdoms.

C. *Naming Organisms*

Our present system for naming organizms can be traced back to the year 1753 when the Swedish botanist, Karl von Linne or "Linnaeus" published *Species Plantarum*. The system is called binomial nomenclature. In it each species is given a two part name, called a scientific name or binomial. The words comprising scientific names are usually derived from Latin or Greek or are words from other languages that are latinized. For example, *Homo sapiens* is the name for humans. *Quercus alba* for white oaks, and *Escherichia coli* for certain bacteria that live in the colon of humans. The first word of the name indicates the genus in which the organism is classified (i.e., *Homo, Quercus, Escherichia*). Notice that the first letter of the genus is capitalized and the word is underlined (or italicized) in print. The second word of the species name is the specific epithet. It begins with a low case letter, is underlined, and is usually descriptive (i.e., "alba" means white; "sapiens" means wise; "coli" refers to the human colon). Note that the word "alba" is not the name for the species of white oaks, *Quercus alba* is. Likewise, the words "coli" and "sapiens" are not the names of species. The scientific name of a species requires the generic name and the specific epithet.

Further, the species name (i.e., *Quercus alba, Escherichia coli*, etc.) is not used in the same sense as the name for an individual human (i.e., John Smith, Jane Doe, etc.). Instead of applying to a single individual, the species name refers to all of the individuals making up a given species.

The complete classifications for humans, white oaks and the colon bacterium are given below.

	Human	**White Oak**	**Colon Bacterium**
Kingdom	Animalia	Plantae	Monera
Phylum	Chordata	Tracheophyta	Schizomycetes
Class	Mammalia	Angiospermae	Eubacteria
Order	Primates	Fagales	Eubacteriales
Family	Hominidae	Fagaceae	Enterobacteriaceae
Genus	*Homo*	*Quercus*	*Escherichia*
Species	*Homo sapiens*	*Quercus alba*	*Escherichia coli*

D. *Identifying Organisms*

In order to aid in identification of organisms, biologists have devised dichotomous keys. A dichotomous key repeatedly gives you two alternatives to choose between. Obtain a plant specimen from your instructor. Examine it and choose between alternatives 1a and 1b.

1a. Plants with needlelike, scalelike or awl-shaped leaves (2)
1b. Plants with broad, flat leaves (3)
    2a. Plants with needlelike leaves *Pinus* sp.
    2b. Plants with scalelike or awl-shaped leaves *Juniperus* sp.

After a choice is made the number in parenthesis to the right of the statement tells you which couplet of alternatives to go to next. For example, if the number (2) appears to the right of the statement, the next alternatives to be considered are 2a and 2b. In a similar way further alternative choices are made until the key leads you to the name of the species or to the name of the group in which the organism is classified.

II. Using a Dichotomous Key

Using the key to some common trees in winter condition, identify twigs of five or more tagged plant specimens provided to you by your instructor. List the number of each specimen and its name on a sheet of paper and turn it in to your instructor to be checked.

KEY TO SOME COMMON LOCAL TREES IN WINTER CONDITION

1a. Leaves evergreen (2)
1b. Leaves deciduous (5)
    2a. Leaves broad and flat with margin toothed—American Holly (*Ilex opaca*)
    2b. Leaves not broad and flat; margin not toothed (3)
3a. Leaves scale-like, hugging twigs—Red Cedar (*Juniperus virginiana*)
3b. Leaves needle-like, in bundles (4)
    4a. Two leaves per bundle—Virginia Pine (*Pinus virginiana*)
    4b. Five leaves per bundle—White Pine (*Pinus strobus*)
5a. Leaf scars opposite (6)
5b. Leaf scars alternate (7)
    6a. Buds in clusters—Red Maple (*Acer rubrum*)
    6b. Buds not in clusters—Flowering Dogwood (*Cornus florida*)
7a. Stem hairy—Staghorn Sumac (*Rhus typhina*)
7b. Stems not hairy (8)
    8a. Leaf scars more than 5 mm in diameter—Tree of Heaven (*Ailanthus altissima*)
    8b. Leaf scars less than 5 mm in diameter (9)
9a. Terminal buds in clusters—Oaks (*Quercus* sp.)
9b. Terminal buds solitary (10)
10a. Axillary buds more than 5 mm long—Beech (*Fagus grandifolia*)
10b. Axillary buds less than 5 mm long (11)
11a. Twigs reddish brown (12)
11b. Twigs not reddish brown (13)
    12a. Axillary buds fine pointed—Birch (*Betula* sp.)
    12b. Axillary buds not fine pointed—Choke cherry (*Prunus virginiana*)
13a. Twigs green—Sassafras (*Sassafras albidum*)
13b. Twigs gray—Tulip poplar (*Liriodendron tulipifera*)

Another type of key is the pictorial key in which a picture is used to help the user identify the key characters. Use the pictorial key to the fleas found on domestic rats in the southern United States with the slide of the cat flea, *Ctenocephalides felis* to become familiar with the operation of the key. Then identify two or more of the unknown fleas using the key. Note that this key is not dichotomous.

Refer to the labeled diagram of a flea below to learn the name of the anatomical features used to differentiate fleas in the key.

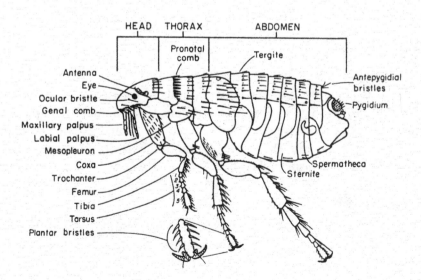

86

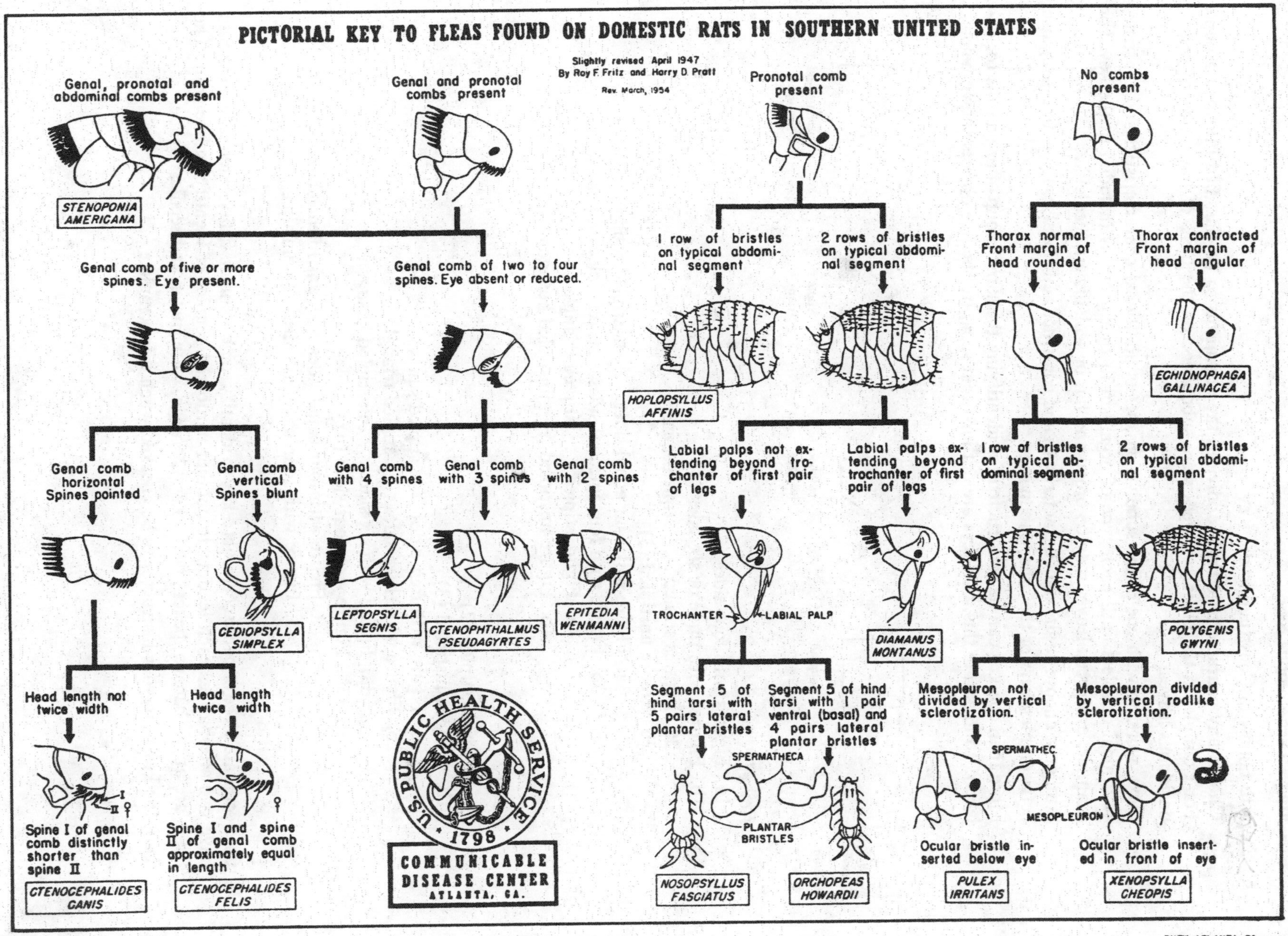
PICTORIAL KEY TO FLEAS FOUND ON DOMESTIC RATS IN SOUTHERN UNITED STATES
Slightly revised April 1947
By Roy F. Fritz and Harry D. Pratt
Rev. March, 1954
Genal, pronotal and abdominal combs present
STENOPONIA AMERICANA
Genal and pronotal combs present
Pronotal comb present
No combs present
Genal comb of five or more spines. Eye present.
Genal comb of two to four spines. Eye absent or reduced.
1 row of bristles on typical abdominal segment
2 rows of bristles on typical abdominal segment
Thorax normal Front margin of head rounded
Thorax contracted Front margin of head angular
HOPLOPSYLLUS AFFINIS
ECHIDNOPHAGA GALLINACEA
Genal comb horizontal Spines pointed
Genal comb vertical Spines blunt
Genal comb with 4 spines
Genal comb with 3 spines
Genal comb with 2 spines
Labial palps not extending beyond trochanter of first pair of legs
Labial palps extending beyond trochanter of first pair of legs
1 row of bristles on typical abdominal segment
2 rows of bristles on typical abdominal segment
CEDIOPSYLLA SIMPLEX
LEPTOPSYLLA SEGNIS
CTENOPHTHALMUS PSEUDAGYRTES
EPITEDIA WENMANNI
TROCHANTER
LABIAL PALP
DIAMANUS MONTANUS
POLYGENIS GWYNI
Head length not twice width
Head length twice width
Segment 5 of hind tarsi with 5 pairs lateral plantar bristles
Segment 5 of hind tarsi with 1 pair ventral (basal) and 4 pairs lateral plantar bristles
Mesopleuron not divided by vertical sclerotization.
Mesopleuron divided by vertical rodlike sclerotization.
I
II
♀
SPERMATHECA
PLANTAR BRISTLES
SPERMATHEC.
MESOPLEURON
Spine I of genal comb distinctly shorter than spine II
Spine I and spine II of genal comb approximately equal in length
Ocular bristle inserted below eye
Ocular bristle inserted in front of eye
CTENOCEPHALIDES CANIS
CTENOCEPHALIDES FELIS
U.S. PUBLIC HEALTH SERVICE
1798
COMMUNICABLE DISEASE CENTER
ATLANTA, GA.
NOSOPSYLLUS FASCIATUS
ORCHOPEAS HOWARDII
PULEX IRRITANS
XENOPSYLLA CHEOPIS
DHEW-ATLANTA, GA.

Figure 8–2.

III. Constructing a Dichotomous Key

A. Steps to Follow in Constructing the Key

    1. Choose any characteristic that subdivides the organisms to be classified into two groups. For example, in the plant key trees were subdivided on the basis of having evergreen or deciduous leaves.

    2. Next, working with one of the subdivisions, select additional characteristics and construct additional couplets until each type of organism in that group is included in the key. For example, evergreen leaves may be broad and flat or scalelike or needlelike.

    3. Similarly, choose additional characteristics and construct additional couplets to distinguish between members of the other subdivision.

    4. More than one key can be constructed for identifying the same group of organisms (or objects). To illustrate this, two keys are outlined that separate the five objects shown below:

| Stipled Square | Unstipled Square | Unstipled Hexagon | Stipled Triangle | Unstipled Trapezoid |

Key #1

    1a. Figures with four sides (2)
    1b. Figures with three or six sides (4)
        2a. Figures stipled . . . stipled square
        2b. Figure unstipled (3)
            3a. Figure with four equal sides . . . unstipled square
            3b. Figure with four unequal sides . . . unstipled trapezoid
                4a. Figure stipled . . . stipled triangle
                4b. Figure unstipled . . . unstipled hexagon

Key #2

    1a. Figures stipled (2)
    1b. Figures unstipled (3)
        2a. Figure with four sides . . . stipled square
        2b. Figure with three sides . . . stipled triangle
            3a. Figure with six sides . . . unstipled hexagon
            3b. Figure with four sides (4)
                4a. Figure with four equal sides . . . unstipled square
                4b. Figure with four unequal sides . . . unstipled trapezoid

B. On a separate piece of paper construct a key to identify some common fruits. Each couplet should have two alternatives each leading to another numbered couplet or identifying one fruit. The fruits are apple, banana, peach, pear, cherry, orange, watermelon and seedless grape.

Turn the key in to your instructor to be checked when it is complete.

Obtain and examine the microscopic slide with a composite of all three shapes of bacteria on it. Make sketches showing the basic shapes.

coccus   bacillus   spirillum

The classification and identification of bacteria is based on their cell shape, size, cell arrangements, presence or absence of capsules, presence or absence and distribution of flagella and endospores, specific staining reactions, colony growth characteristics on various media and on biochemical reactions.

Observe and make sketches illustrating capsules, endospores and flagella.

**Encapsulated   Endospore-forming   Flagellated
Bacterium   Bacterium   Bacterium**

Within 24 to 48 hours the growth and multiplication of a single bacterium on a solid culture medium can produce a macroscopic visible growth called a colony. Examine and note differences in the colonies of representative bacteria growing on tryptic soy agar.

bacterium   notes

_____   _____

                            _____

_____   _____

                            _____

_____   _____

1. Smear Preparation
   a. With the flat end of a clean toothpick gently scrape some cells from the inside surface of your cheek or from in between your teeth.
   b. Spread the material out in the center of a clean glass slide.
   c. Allow the preparation to air dry.
   d. While holding one end of the slide between two fingers, pass the smear through a flame two or three times. This is called heat fixing and serves to kill the bacteria and causes them to adhere to the slide.

2. Simple Stain

   a. Cover the smear with methylene blue dye for one minute.

   b. Rinse the dye off with a few drops of water and carefully blot the preparation between the two halves of a folded paper towel.

   c. Examine the stained slide under the oil immersion objective and make drawings of your observations.

B. Cyanobacteria (Blue-Green Bacteria)

Blue-green bacteria are unicellular or colonial with cells joined by a gelatinous sheath. They contain chlorophyll a but not in chloroplasts and are photosynthetic. Reproduction is asexual by fission or fragmentation of colonies or by sexual conjugation on occasion. Some of them, along with other bacteria, can fix atmospheric nitrogen.

Prepare a wet mount of each of the three live cultures listed below and make a sketch of each.

Gloeocapsa　　　　　　　　　　*Nostoc*　　　　　　　　　　*Oscillatoria*

Which of the above:

1. Forms a long chain of cells with an occasional large clear cell, heterocyst, present?

   _____

2. Forms spherical cells singly, in pairs or in small groups, each surrounded by a thick gelatinous sheath? _____

3. Forms chains of uniform cells? _____

Check your answers to the above questions by observing prepared slides of each species.

II. Kingdom Protista

Representatives have eukaryotic cells and are unicellular, colonial or if multicellular remain relatively undifferentiated. Cells, if motile, move by 9 + 2 cilia or flagella or by pseudopodia. Forms of nutrition include absorption, ingestion and photosynthesis. The group includes algae, protozoans and slime molds. Flagellated protistans are thought to be the ancestors of animals and plants. A number of the groups included in Kingdom Protista have been elevated to kingdom status in newer classification schemes.

A. *Eukaryotic Algae*

The eucaryotic algae are a diverse assemblage. They vary in shape from a more or less spherical single cell to a ball of cells, a filament of cells, a sheet of cells to various branching forms. Some, the giant kelps, attain lengths of over 30m. Even then they remain relatively undifferentiated. Being nonvascular they do not form true roots, stems or leaves. Algae are extremely important in ecosystems as producers. Most of them are aquatic (streams, ponds, oceans) but some are terrestrial (moist soil, rocks, bark of trees). Algae are classified according to cell structure, pigments, food storage compounds, number of flagella and mode of reproduction.

1. Green Algae

The green algae are of particular interest because they appear to be on the line of evolution that led to the higher plants. In fact, some biologists now classify them in the plant kingdom.

The modern green alga *Chlamydomonas* may represent a relatively unmodified descendent of early aquatic eukaryotes. *Chlamydomonas* reproduces asexually by mitotic division of the nucleus followed by cytokinesis. Two, four, . . ., 16 or more cells may be produced that develop into motile cells (called *zoospores*). Each zoospore is released and becomes a new *Chlamydomonas*.

In sexual reproduction haploid *gametes* are produced inside each *Chlamydomonas* cell. The cell is called a gametangium, a one-cell structure in which gametes are produced. Gametes released from different cells swim about. Eventually two identical gametes, called *isogametes*, formed by different cells (+ and − strains) unite to form a flagellated, diploid zygote. The four flagella are soon lost and the zygote develops a thick wall and becomes dormant.

Under favorable conditions the resistant zygote undergoes meiosis to produce four haploid cells that are released and give rise to new *Chlamydomonas*. Before continuing the exercise:

a. Prepare and examine a wet mount of living *Chlamydomonas*. Referring to Figure 9–1, locate the following structures—large cup-shaped chloroplast, stigma (red eyespot for detecting light), pyrenoid (starch forming body), pair of anterior flagella and cell wall.

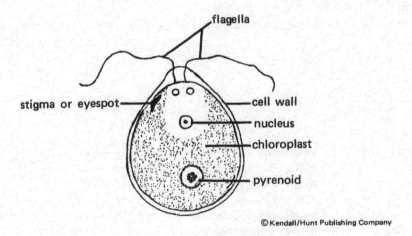

Figure 9–1. Chlamydomonas.

b. A sexual cycle in which the zygote is the only diploid cell is referred to as a *haplontic sexual cycle*. Using *Chlamydomonas* as a model of a primitive green alga, the early green algae probably were:

    (1) unicelluar,

    (2) flagellated,

    (3) photosynthetic,

    (4) compartmented—specific functions occurring within organelles,

    (5) capable of reproducing asexually by forming flagellated zoospores and sexually by the union of flagellated isogametes to form a diploid zygote,

    (6) aquatic organisms with a haplontic sexual cycle.

Use the space below to outline the asexual and sexual reproduction in *Chlamydomonas*.

Evolutionary Trends

Starting with a chlamydomonaslike organism, there are three trends that parallel each other and culminate in the higher plants.

    a. An increase in the complexity of the vegetative body, and

    b. A trend toward increasing difference between male and female gametes and/or male and female reproductive organs.

    c. A trend toward increased emphasis on diploid stages in the life cycle.

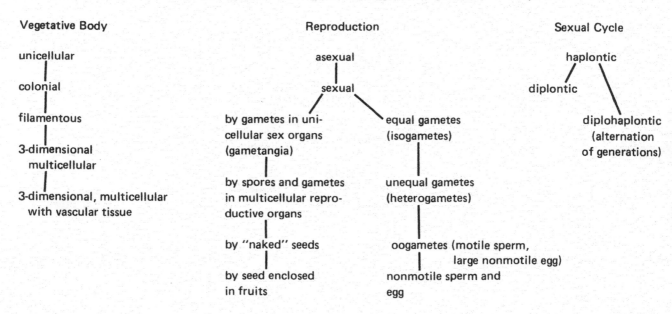

Vegetative Body	Reproduction	Sexual Cycle
unicellular	asexual	haplontic
colonial	sexual	diplontic
filamentous	by gametes in unicellular sex organs (gametangia) equal gametes (isogametes)	diplohaplontic (alternation of generations)
3-dimensional multicellular	by spores and gametes in multicellular reproductive organs unequal gametes (heterogametes)	
3-dimensional, multicellular with vascular tissue	by "naked" seeds oogametes (motile sperm, large nonmotile egg)	
	by seed enclosed in fruits nonmotile sperm and egg	

At this point, you have already examined isogamy in *Chlamydomonas*. In *Spirogyra* (Fig. 9–2), sexual reproduction consists of filaments of opposite mating types lining up adjacent to each other and forming small projections which grow toward each other from cells of the opposing filaments. The projections eventually join, and the end walls dissolve forming a conjugation tube between the two cells. The protoplasts of the conjugating cells act as gametes. Since the gametes are similar in appearance and size, what

kind of gametes are they? _____

The contents of one cell migrates through the tube and unites with the contents of the opposite cell to produce a zygote which matures to form a resistant zygospore. A single haploid filament develops from the zygospore following meiosis.

Examine a prepared slide of *Spirogyra* showing conjugation. Identify *conjugation tubes*, *migrating isogametes*, and *zygotes*.

If fresh material is available, prepare and examine a wet mount of *Spirogyra*. Add a drop of iodine solution at one edge of the coverslip to test for the presence of starch.

Did you get a positive test for starch? _____

If so, where was the starch located?_____

Make your own sketches of *Spirogyra* in the space to the right of Figure 9–2.

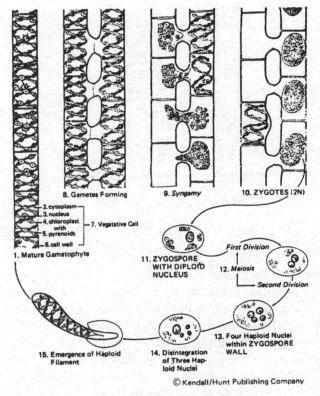

Figure 9–2. Life cycle of Spirogyra.

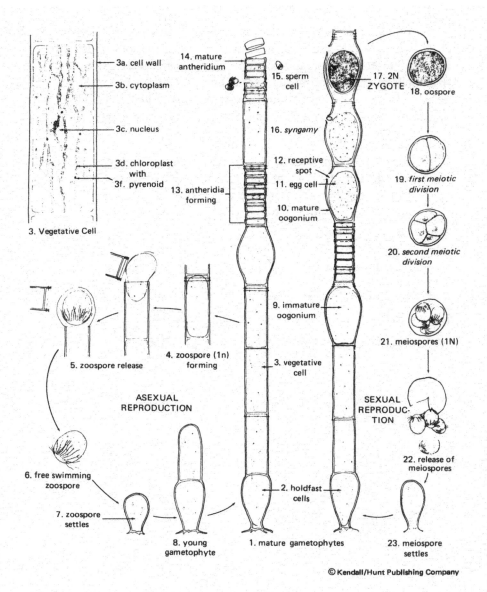

The following labels appear on the figure:

3a. cell wall
3b. cytoplasm
3c. nucleus
3d. chloroplast with
3f. pyrenoid
3. Vegetative Cell

14. mature antheridium
15. sperm cell
16. syngamy
12. receptive spot
11. egg cell
10. mature oogonium
13. antheridia forming
9. immature oogonium
3. vegetative cell

ASEXUAL REPRODUCTION

4. zoospore (1n) forming
5. zoospore release
6. free swimming zoospore
7. zoospore settles
8. young gametophyte
1. mature gametophytes
2. holdfast cells

17. 2N ZYGOTE
18. oospore
19. first meiotic division
20. second meiotic division
21. meiospores (1N)
22. release of meiospores
23. meiospore settles

SEXUAL REPRODUCTION

© Kendall/Hunt Publishing Company

Figure 9-3. Life cycle of *Oedogonium*.

Next, refer to the diagram of the life cycle of *Oedogonium* (Fig. 9–3) a filamentous, freshwater green alga. *Oedogonium* reproduces asexually by forming multiflagellated zoospores which attach to a substrate and give rise to a filament. Sexual reproduction involves the formation of two multiflagellated sperm in each *antheridium* (small cell) and a large, nonmotile egg formed in an *oogonium* (enlarged cell). A motile sperm swims to the oogonium and enters it through a pore to fertilize the egg. The resulting diploid zygote undergoes meiosis to form haploid meiospores. Since the zygote is the only diploid cell, what type of sexual cycle (haplontic, diplontic or diplohaplontic) does *Oedogonium*

have? _____

Alternation of Generations (Diplohaplontic Sexual Cycle)

Many biologists believe that the green algal ancestors of higher plants had a life cycle which included an alternation of two independent phases (or generations) which were similar in size. In this type of life cycle, gametes produced by a haploid gametophyte generation unite in the process of fertilization to form a diploid zygote. Subsequent de-

96

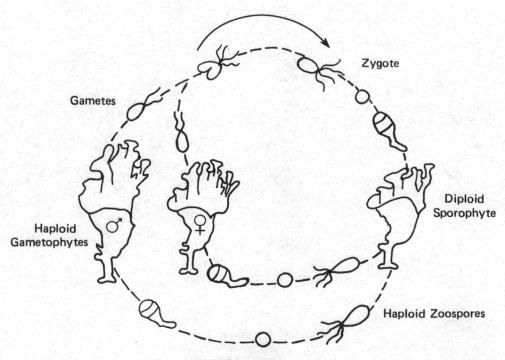

Figure 9-4. Life cycle of *Ulva*.

velopment of the zygote gives rise to a diploid sporophyte generation capable of producing haploid spores by meiosis. Development of the haploid spores gives rise to new haploid gametophytes. The life cycle of the sea lettuce, *Ulva* Figure 9-4, is an example of the diplohaplontic cycle.

Examine preserved or fresh material of *Ulva*. How does the plant body of *Ulva* differ

from the plant bodies of other green algae examined previously? _____

2. Brown Algae (Figure 9-5, A-C)

Brown algae are the most complex algae structurally and range from simple filaments to massive vegetative bodies. Pigments include chlorophylls a and c and fucoxanthin, a brown pigment. Food is stored as an oil, or as a carbohydrate, laminarin. Both sexual and asexual reproduction occur. Alginates, extracted from the walls of some brown algae, are used as stabilizers in ice cream, sherbets and cream cheeses, and as a gelling agent in milk puddings. Representatives include the kelp *Laminaria,* the rockweed *Fucus,* and the sargasso weed, *Sargassum.*

Examine specimens as available.

3. Red Algae (Figure 9-5, D and E)

Most organisms in this group are multicellular. Pigments include chlorophylls a and d, and the biliprotein phycoerythrin, which is red. Food is stored as Floridean starch. Sexual and asexual reproduction occur. Motile cells are not produced. One important use of red algae is in the production of agar used as a gelling agent in microbiological media and in fruit pies. Another extract, carrageenan, is used as a suspending agent in chocolate milk and as a gelling agent in milk puddings and pie fillings. Representatives include *Chondrus* and *Polysiphonia.*

4. Dinoflagellates (Figure 9-6, A)

The dinoflagellates may be motile or nonmotile, unicells or filamentous. Photosynthetic pigments include chlorophylls a and c, diadinoxanthin and dinoxanthin. Food is stored as starch. The most conspicuous distinguishing attribute of dinoflagellates is the pres-

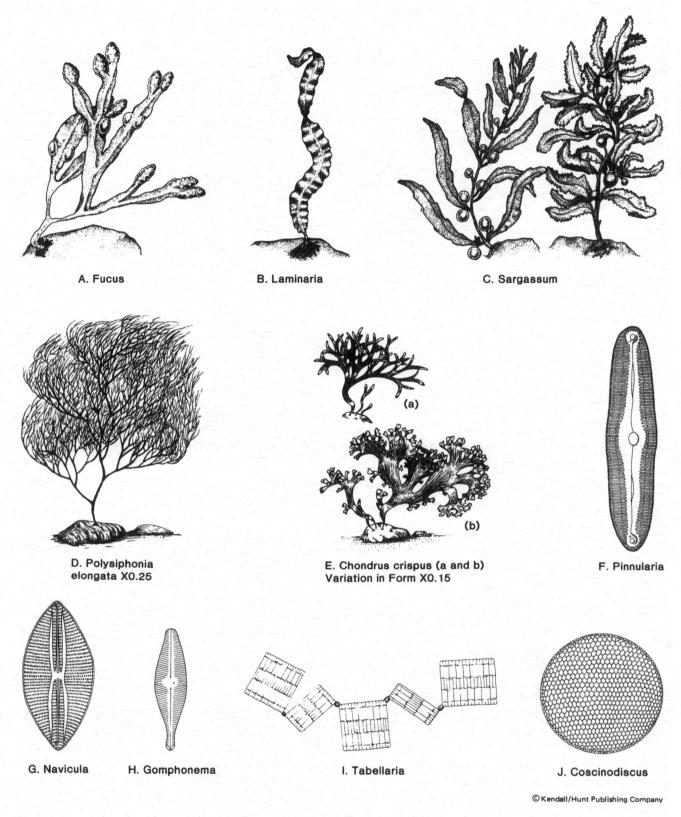

A. Fucus

B. Laminaria

C. Sargassum

D. Polysiphonia
elongata X0.25

E. Chondrus crispus (a and b)
Variation in Form X0.15

(a)

(b)

F. Pinnularia

G. Navicula

H. Gomphonema

I. Tabellaria

J. Coscinodiscus

Figure 9–5. Representative protists.

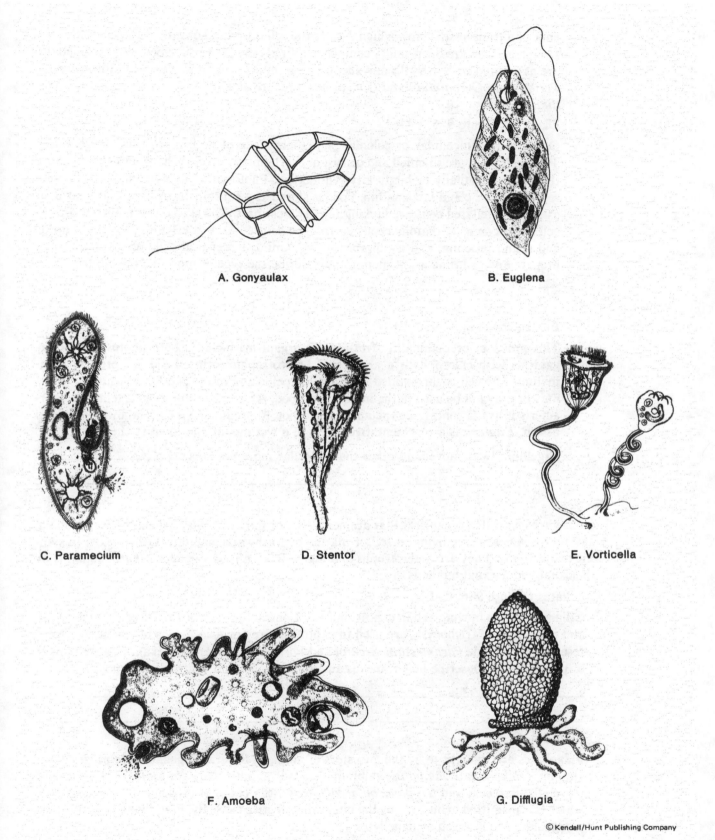

A. Gonyaulax

B. Euglena

C. Paramecium

D. Stentor

E. Vorticella

F. Amoeba

G. Difflugia

Figure 9-6. Representative protists.

ence of a transverse furrow which divides the cell into equal or rather unequal semicells. Dinoflagellates are the primary component of red tides. Toxins formed by dinoflagellates are associated with fish kills and shellfish poisoning in humans. Representatives include *Peridinium, Gyrodinium* and *Gonyaulax*. Prepare and examine a wet mount of *Gyrodinium*.

5. Diatoms (Figure 9–5, F–J)

Diatoms are unicellular or colonial planktonic algae of both marine and fresh water. Their cell walls are silicified and consist of overlapping halves. When viewed from above the cells are radially or bilaterally symmetical. Photosynthetic pigments include chlorophylls a and c and fucoxanthin. Food is stored as a carbohydrate, chrysolaminarin, or as oils. Flagellated cells are lacking for the most part. The siliceous remains of diatom cell walls comprise diatomaceous earth. This is used as a filtering agent, in insulation and sound proofing, and as an abrasive. Examine a prepared slide of diatoms. Next, prepare and examine a wet mount and describe the type of movement exhibited.

6. Euglenoids (Figure 9–6, B)

This group of predominantly fresh water organisms includes both photosynthetic organisms with chlorophylls a and b and heterotrophic forms. Food is stored as paramylum. The cell is surrounded by a flexible pellicle in most species. An anterior gullet may or may not be used in the ingestion of food. At least one anterior flagellum is characteristic as is a contractile vacuole and eyespot. Reproduction is asexual by longitudinal fission. *Euglena* is a representative. Prepare a wet mount and identify the eyespot and

flagellum. Does movement other than flagellar movement occur? _____

_____

B. Protozoa

Protozoa are one-celled organisms that are nonmotile or move by cilia, flagella or pseudopodia. Many protozoans are free-living in soil or marine or freshwater habitats. Others exist in symbiotic associations. They are heterotrophs and digest food in food vacuoles. Reproduction is by both sexual and asexual methods.

1. Ciliates (Figure 9–6, C–E)

Cilia are used in moving and in capturing food. Contractile vacuoles are used to accumulate and expel water. Trichocysts are used in capturing prey or as a means of protection. Representatives include *Paramecium, Stentor* and *Vorticella*. Prepare and examine a wet mount of paramecium. Describe how it moves and identify cilia and contractile vacuoles.

_____

_____

2. Flagellates

Flagellates are thought to be the ancestors of animals. Several flagellates are parasites including *Trichomonas vaginalis,* which may cause vaginitis, *Giardia lamblia,* the cause of traveler's diarrhea, and *Trypanosoma gambiense,* the cause of African sleeping sickness, which is transmitted to humans by the bite of an infective tsetse fly. Free-living heterotrophic euglenoids are classified as flagellates.

Examine a blood smear showing trypanosomes. Sketch a trypanosome from the plasma.

3. Rhizopods (Figure 9–6, F and G)

Pseudopodia are used in locomotion and food getting. Food vacuoles and contractile vacuoles are formed. Some form shells. *Entamoeba histolytica* causes amoebic dysentery. Others including *Amoeba proteus* and *Difflugia* are free living.

Prepare a wet mount of an amoeba and sketch and describe how it moves.

_____

_____

Slime molds have ameboid stages but will not be studied in this exercise.

4. Apicomplexans

All apicomplexans are parasites. A representative of this group will be studied in the exercise on symbiosis.

III. Kingdom Fungi

The term "fungi" is a general term applied to several diverse groups that are nonvascular, achlorophyllous (and therefore heterotrophic), absorptive, eukaryotic, and that reproduce by forming spores (sexual and/or asexual). Fungi are unicellular or multicellular. Multicellular fungi are usually composed of filaments called hyphae. The entire body formed of hyphae is called a mycelium. Since fungi exist as decomposers or parasites they have coevolved with other groups.

A. Zygomycetes (Phylum Zygomycota)

Rhizopus (Figure 9–7) is one of the most common fungi and may be found growing on many foods (i.e., bread). Examine a mature culture of *Rhizopus* with a dissecting microscope. The mycelium is differentiated into three kinds of hyphae: (1) stolons—long hyphae that grow over the surface of the substrate, (2) rhizoids—rootlike hyphae that penetrate the substrate, and (3) sporangiophores—upright hyphae that bear globular sporangia at their tips. Using forceps, remove a small portion of the mycelium from a plate culture and mount it in a drop of water on a slide. Identify stolons, rhizoids, sporangiophores, sporangia and spores. Are the spores motile

or nonmotile? _____ Can you see cross walls in the hyphae? _____

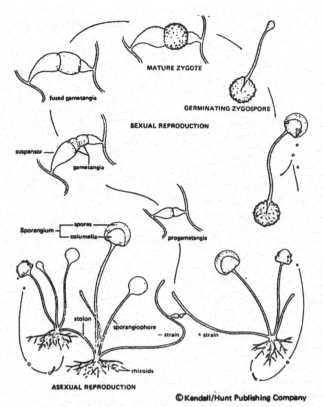

Figure 9–7. Life cycle of *Rhizopus*.

© Kendall/Hunt Publishing Company

What color do the sporangiospores appear in the wet mount? _____ In mass the spores appear black. When the sporangium breaks open the nonmotile spores are wind dispersed. On germination the spores give rise to hyphae without cross walls. Sexual reproduction (conjugation) of *Rhizopus* requires two mating types (+ and −). When the two mating types grow in close contact small projections grow toward each other from the opposite hyphae. Just behind the contacting tips cross walls form which cut off terminal cells (gametangia) that act as gametes. A zygospore (sometimes called a zygote) forms between the opposite filaments.

Obtain and examine a prepared slide of *Rhizopus* conjugation and identify representative stages of conjugation.

B. *The Sac Fungi or Ascomycetes* (Phylum Ascomycota)

The Sac Fungi or Ascomycetes are a large group of economically important fungi that characteristically produce a sac or ascus (plural is asci) during sexual reproduction in which haploid spores (ascospores) are formed following meiosis. The hyphae of multicellular representatives are septate (have cross walls) in contrast to those of zygomycetes. Many representatives also produce asexual spores or conidia that are not contained in a sporangium. No motile cells are formed in the group.

1. *Peziza* (Fig. 9–8)

In *Peziza* the asci are borne on the upper surface of a cup-shaped fruiting body.

Examine a prepared slide of *Peziza* and identify fruiting body, asci and ascospores. How many ascospores are contained in a single ascus?

ascospore

ascus

fruiting bodies

Figure 9-8. Cup fungus—*Peziza*.

2. *Penicillium* (Fig. 9–9)

Members of this genus are important in food spoilage, cheese production, penicillin production, etc. The genus name, *Penicillium*, refers only to the imperfect form (asexual stages) of this fungus. The perfect form (sexual stages), which will not be studied in this exercise, places this fungus in the ascomycetes.

Examine a prepared slide of *Penicillium* showing conidia borne on specialized hyphae called conidiophores. Identify conidia and conidiophores.

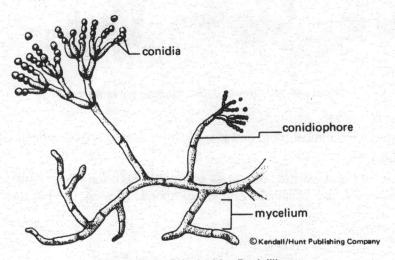

conidia

conidiophore

mycelium

© Kendall/Hunt Publishing Company

Figure 9-9. Blue mold—*Penicillium*.

3. *Saccharomyces cerevisiae*

Strains of this species of yeast are used in the baking and brewing industries because of their ability to ferment sugars to produce carbon dioxide and ethyl alcohol.

Under certain conditions sexual processes give rise to ascospores in an ascus. Asexual reproduction occurs by *budding*. Literally, a blowout (bud) occurs at some point on the parent cell. Mitosis occurs and one of the daughter nuclei migrates into the bud. Subsequently, the bud enlarges, is walled off and eventually separates from the parent cell. Short chains of cells may be produced by rapid budding.

Obtain a drop of yeast suspension, prepare a wet mount and locate and sketch budding yeast cells.

103

C. *The Club Fungi or Basidiomycetes* (Phylum Basidiomycota)

The club fungi or Basidiomycetes include many of the more conspicuous fungi such as mushrooms, toadstools, bracket or shelf fungi and puffballs as well as the rusts and smuts which are obligate parasites on many vital economic plants.

In sexual reproduction, haploid spores (basidiospores) are produced following meiosis on the outside of a club-shaped structure called a basidium (plural is basidia). The hyphae, like those of ascomycetes, are septate. Some basidiomycetes also produce asexual spores called conidia. No motile cells are formed in this group.

1. *Agaricus*

   Mushrooms form an extensive vegetative mycelium beneath the ground. The more obvious part of the life cycle is the above ground fruiting body or mushroom. The basidia are borne on the surface of gills located on the underside of a mushroom cap or pileus. In a mature mushroom the pileus is elevated above the ground by a stalk or stipe. A ring or annulus usually surrounds the stipe.

Figure 9–10. Growth habit. Agaricus type mushroom.

© Kendall/Hunt Publishing Compai

   Examine a fresh mushroom and identify stipe, annulus (if present), gills and pileus.

2. *Coprinus*

   Examine a prepared slide of a cross section through the pileus (cap) of *Coprinus*. Identify and make a sketch of basidia and basidiospores. See Fig. 9–11 A and B.

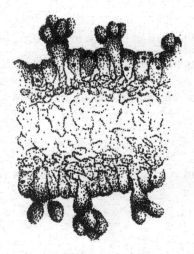

A. Section of a gill.

B. Basidium bearing basidiospores.

Figure 9–11.

D. *Fungi Imperfecti* (Deuteromycetes)

For convenience sake the imperfect (asexual) stages of many fungi with known sexual stages, mostly ascomycetes, and the imperfect stages of fungi with no known sexual stages are classified in this group. Such a system provides a means of identifying a fungus based on imperfect stages when they are encountered. This is of paramount importance when the fungus is associated with human disease or is of economic importance in agriculture or industry.

What representatives of this group have you already examined in this exercise?

REVIEW QUESTIONS

1. The three basic shapes of bacteria are:

   a. _____ b. _____ c. _____

2. Explain the purpose of heat fixing a bacterial smear.

3. Identify the major trends that are believed to have occurred in the evolution of algae and plants.

4. Which of the following green algae form oogametes? (*Chlamydomonas, Spirogyra, Oedogonium*)

5. Identify the algal group associated with each of the following:

   a. Source of the agar used in microbiological media _____

   b. Primary component of red tides _____

c. Their silicified cell walls consist of overlapping halves _____

d. Members of this group are associated with fish kills and shellfish poisoning in humans _____

e. Some representatives have massive vegetative bodies, some are the source of alginates _____

6. Identify three types of movement used by protozoans and give an example of an organism that uses each type.

a. _____   b. _____   c. _____

7. The body of a fungus is called a _____ and consists of threadlike filaments called

_____ .

Match each organism below with the correct type of spore:

____ 8. *Coprinus*   a. ascospore

____ 9. *Penicillium*   b. basidiospore

____ 10. *Peziza*   c. conidium

____ 11. *Rhizopus*   d. zygospore

12. Which spore type on the right above is an asexual spore? _____

13. Which fungus on the left above reproduces asexually by forming sporangiospores? _____

EVOLUTION OF PLANTS

Objectives

Upon completion of this exercise you should be able to:

1. Cite at least four reasons for a possible green alga ancestry of land plants.

2. Outline the basic alternation of generations life cycle of a land plant and indicate the genetic complement (haploid, diploid) of each phase.

3. Identify the three major groups of land plants and recognize representative specimens from each group.

4. Trace the life cycle of designated specimens from each of the three major groups of land plants.

5. Identify the following from specimens, models, drawings and prepared slides: gametophyte, archegonium, egg, antheridium, sperm, sporophyte, sporangium, and spore.

6. For each specimen examined, describe its adaptations for life on land and its limitations for land existence.

7. Identify and distinguish among the following from specimens, models, drawings and prepared slides: thallus, stomata, cuticle, rhizoid, protonema, prothallus, sorus, staminate cone, ovulate cone, pollen, pistil, and stamen.

8. Trace the evolutionary trends in land plants in reference to:

   a. the comparative dominance of the gametophyte and sporophyte generations
   b. the means of species dispersal by spore and seed.

9. Answer all questions in the exercise.

I. Introduction

Two evolutionary lines are evident among the first land plants. One line of evolution led to plants in which no specialized supporting and conducting cells are formed. These plants remained small and inconspicuous and are restricted to moist, shady habitats for the most part. They include the mosses, liverworts and relatives which are collectively known as *bryophytes*.

Plants in the second line, *vascular plants* developed specialized conducting and supporting tissues which, along with other adaptations, enabled them to exploit the landscape to its fullest. Both groups are thought to have arisen from green algae.

Support for green algal ancestry is based on common characteristics possessed by certain green algae and the land plants. Both groups have: the same type of chlorophyll molecule, cell walls of cellulose, and store food in the form of starch. In addition, the alternation of generations (diplohaplontic cycle) exhibited by certain green algae also is present in the life cycle of land plants.

II. The Alternation of Generations and Evolutionary Trends in Land Plants

A tremendous amount of variation exists among the land plants, but understanding the basic alternation of generations life cycle makes it easier to trace the evolutionary development of the land plants. The basic plant life cycle (Fig. 10–1) alternates between a gamete producing phase (the GAMETOPHYTE GENERATION) and a spore producing phase (the SPOROPHYTE GENERATION). The gametophyte generation consists of a haploid multicellular plant body. The gametophyte produces gametes (eggs and sperm) in special sex organs called archegonia and antheridia.

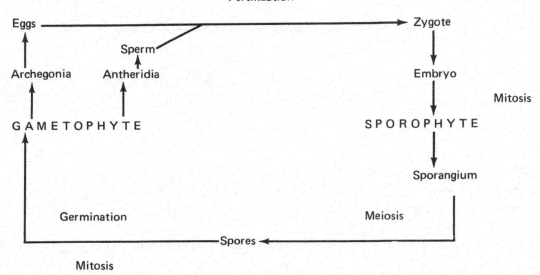

Figure 10–1. Basic life cycle of a land plant.

Based on the information in Fig. 10–1, the archegonia produce _____ and the antheridia produce _____ . The gametes produced by the gametophyte unite to form a diploid cell called the zygote. The zygote undergoes cell division (mitosis) to produce a diploid multicellular body known as the sporophyte. A special chamber (the sporangium) in the sporophyte is the site of spore production. Spores are produced by the process of meiosis. Consequently, spores are haploid. Spores provide a means of dispersal in simple land plants and ultimately give rise to another gametophyte.

Your instructor will assign several specimens to examine representing three major groups of land plants:

1. Bryophytes—no specialized conducting (vascular) tissue
   —swimming sperm
   —dispersal by spores

2. Seedless vascular plants—vascular tissue present
   —swimming sperm
   —dispersal by spores

3. Seed plants—vascular tissue present
   —pollen produced
   —dispersal by seeds

For each specimen you examine: (see chart at end of exercise)

1. Write a brief description or make a sketch of:
   a. the gametophyte
   b. the archegonia and antheridia
   c. the sporophyte (including its relationship to the gametophyte)
   d. the sporangia

2. Explain how the sperm travels to the egg.

3. Indicate which generation (gametophyte or sporophyte) is the dominant generation (i.e. larger, more independent, etc.).

4. Describe adaptations for life on land.

III. Bryophytes (Liverworts, Mosses, Hornworts)

Bryophytes are small plants and most require a moist environment. Their size limitation is due to the absence of vascular tissue (specialized supporting and conducting tissue). Their restriction to a moist habitat is due to the requirement that sperm must swim to the egg. Bryophytes, however, do possess adaptations for life on land.

1. Liverworts (Hepatopsida) (Use Fig. 10–2 as a general guide to the identification of the various stages in the life cycle of a liverwort.)

   Marchantia will be used to illustrate the basic life cycle of liverworts. The *gametophyte* plant is the small flattened unspecialized body called a thallus. The lobed shape of the thallus resembles the shape of the human liver, hence the name, liverworts. About 1500 species of liverworts possess flattened bodies similar to *Marchantia*. The remaining 6500 species have bodies that are leafy in appearance.

   Obtain and examine a slide of *Marchantia* thallus c.s. The thallus displays several adaptive features for life on land such as an outer (upper) covering of cells called an *epidermis* and a *waxy cuticle* which is secreted by the epidermis.

   In what way are these features adaptations for life on land?

   _____

   In addition to preventing water loss, land plants must be able to obtain carbon dioxide gas for photosynthesis and release excess oxygen. Identify the small pores (*stomata*) in the epidermis which allow for gas exchange. Examine the lower surface of the thallus. Notice the small cellular extensions called *rhizoids*. Although they are not true roots by any means, what functions do

   they accomplish? _____

   In addition to providing anchorage in the soil, the rhizoids also aid in the absorption of water and minerals.

   Sketch a section of the liverwort gametophyte (thallus) and label cuticle, epidermis, stomata and rhizoids.

   If available examine a thallus of a fresh liverwort.

   The reproductive function of the gametophyte is to produce *gametes*. *Marchantia* is heterothallic. This means that one type of gametophyte produces only male gametes and another type produces female gametes. Examine specimens or models of *Marchantia* gametophytes.

   One type of gametophyte possesses stalk-like structures which resemble umbrellas. The arched lobes at the top are the archegonial heads which contain the archegonia on their under surfaces. The second type of gametophyte possesses stalk-like structures which are flattened at the top. They are the antheridial heads and bear antheridia on the upper surface. The multicellular archegonia and antheridia represent advances in reproduction over the green algae in which the gametes are produced in single cells.

   Using Fig. 10–2 as a guide, identify archegonia and antheridia on prepared slides labeled *Marchantia* archegoniophore and *Marchantia* antheridiophore respectively. In the space below diagram an archegonium and an antheridium.

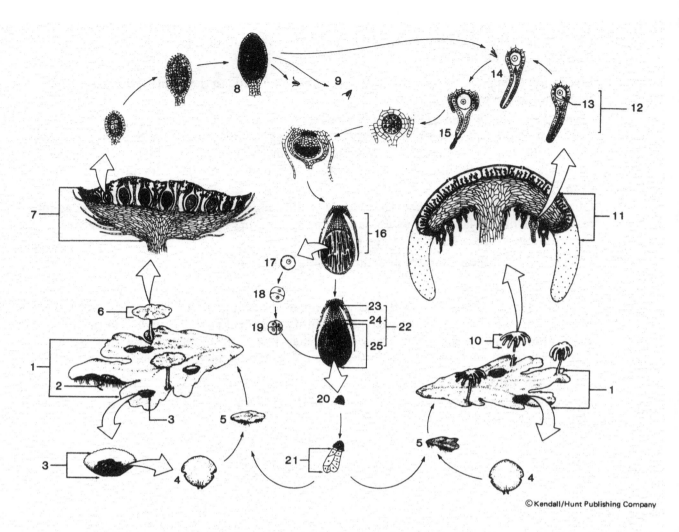

1. Mature Gametophyte
2. Rhizoids
3. Cupule Containing Gemmae
4-5. Asexual Reproduction
   4. Gemma Released from Cupule
   5. Young Gametophyte Thallus
6. Antheridial Head
7. Antheridial Head (L.S.)
8. Mature Antheridium
9. Sperm
10. Archegonial Head
11. Archegonial Head (L.S.)
12. Archegonium (L.S.)
   13. Egg Cell

14. Fertilization
15. Archegonium with Zygote
16. Sporophyte Maturing
17. Spore Mother Cell
18-19. Meiosis
   18. Dyad
   19. Tetrad of Meiospores
20. Meiospore Released
21. Young Gametophyte Thallus
22. Sporophyte
   23. Foot
   24. Stalk
   25. Capsule

Figure 10–2. Life cycle of *Marchantia*.

The sperm of bryophytes are flagellated and require water in which to swim to the egg. Notice how the top of the antheridial head is flattened so that rain drops will splash on them and carry the sperm to the archegonial heads. Note that the umbrella-like prongs of the archegonial head are well adapted to hold the splashed drops of water. When a haploid sperm fertilizes a haploid egg, a diploid zygote is produced.

The diploid zygote undergoes cell division to form a multicellular diploid *embryo,* which is located within the archegonium. The embryo continues to develop into a multicellular diploid *sporophyte.* Examine a model or preserved specimen with sporophytes attached on the underside of the archegonial head. Note the size of the sporophyte in relation to the size of the gametophyte. Examine the prepared slide labeled *Marchantia* sporophyte median l.s. and identify three distinct regions. Some cells of the sporophyte have invaded the gametophyte tissue and serve to absorb nutrients from the gametophyte. Just below these "foot cells," is another group of cells which form a short thick stalk. Below the stalk, the remaining sporophyte cells form a capsule. Inside the capsule, diploid cells are undergoing the process of meiosis in which they will give rise to haploid *meiospores.* Each diploid cell undergoing meiosis produces four haploid meiospores. The haploid meiospores are released when the enlarged capsule dries out and ruptures. Wind dispersal of the spores is aided by slender threads within the capsule called elaters. As the elaters dry they twist and coil. The jerking movements of the elaters aid in exposing the mass of spores to the wind. In the space below diagram a sporophyte and label meiospores.

If a spore falls on suitable soil, it will germinate and give rise to a gametophyte.

You may have noticed the presence of small cup-like structures on the dorsal surface of the *Marchantia* gametophyte. These contain specialized structures called *gemmae.* The gemmae are the means by which *Marchantia* reproduces asexually.

In summary, the alternation of generations life cycle of the liverwort, *Marchantia,* consists of a flattened lobed haploid _____ which produces gametes by means of specialized multicellular sex organs called _____ and _____ . The motile sperm require the presence of water in order to swim to the egg. The union of the haploid sperm and haploid egg results in the formation of a diploid _____. Still within the archegonium, the zygote develops into an embryo which in turn develops into a _____ . Within the capsule of the sporophyte diploid cells undergo the process of _____ and produce haploid _____ . The haploid meiospores are released and eventually germinate. They then undergo cell division to form the multicellular haploid _____ .

2. Mosses (Bryopsida) Fig. 10-3

The individual moss plant consists of a slender leafy axis, either erect or prostrate, which may or may not have multicellular absorptive branches called rhizoids. (Note—true roots, stems and leaves are lacking for by definition, vascular tissue is present in true roots, stems and leaves.)

It is characteristic of mosses that they rarely occur as individuals but form extensive groups or colonies on moist soil, rocks, and wood. They are widespread in rainy and humid places. Sexual reproduction is oogamous.

Gametophyte—mature stage. Examine a living "leafy" moss gametophyte. This is the mature gametophyte which is photosynthetic. Identify "leaves," "stems," and rhizoids. Remove one gametophyte from the moss colony and observe it with the stereoscopic microscope. Note that the

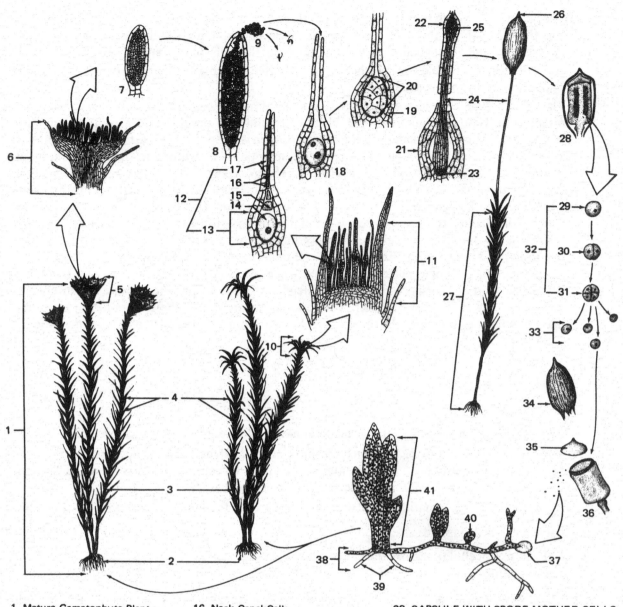

1. Mature Gametophyte Plant
2. Rhizoids
3. Non-Vascular "Stem"
4. Non-Vascular "Leaves"
5. Antheridial Head
6. Antheridial Head (L.S.)
7. Maturing Antheridium
8. Mature Antheridium
9. Sperm
10. Archegonial Head
11. Archegonial Head (L.S.)
12. Mature Archegonium
   13. Venter
   14. Egg Cell
   15. Ventral Canal Cell

16. Neck Canal Cells
17. Neck Cells
18. ZYGOTE Formed in Fertilization
19. Archegonium Venter Wall
20. EARLY SPOROPHYTE
21. Enlarged Venter Wall
22. Neck Cells Developing into Calyptra
23-25. YOUNG SPOROPHYTE
   23. FOOT
   24. STALK
   25. CAPSULE
26. Calyptra Covering CAPSULE
27. Female Gametophyte with Attached
Maturing SPOROPHYTE

28. CAPSULE WITH SPORE MOTHER CELLS
29. SPORE MOTHER CELL
30. Dyad
31. Tetrad of Meiospores
32. Meiosis
33. Meiospore
34. Calyptra Removed from Capsule
35. Capsule Lid
36. Meiospores Released from Capsule
37. Germinated Meiospore
38. Protonema
39. Rhizoids
40. Bulbil ("Bud")
41. Young Gametophyte Plant

Figure 10–3. Life cycle of a moss.

"leaves" do not have specialized vascular tissue (i.e. veins are absent). The gametophyte contains special sex organs in which the gametes are produced. Many male gametes or sperm are produced in oval shaped antheridia and the individual female gametes or eggs are produced in the enlarged base of each slender archegonium (plural = archegonia). Examine the prepared slides of moss antheridia and archegonia and in the space below make a diagram of one antheridium and one archegonium. Label sperm and egg.

You may have noticed that the top (or splash cup) of the male gametophyte is rather flat compared to the leafy female gametophyte. In what way does the difference in shape aid in the transfer of sperm to the archegonium when it rains?

The sperm are motile and require a film of water to move from antheridium to archegonium and down the neck of the archegonium. The fertilized egg or zygote develops into a sporophyte by numerous cell divisions and morphological changes.

Moss sporophyte. Examine living specimens which consist of both gametophyte and sporophyte generations. The sporophyte consists of a terminal capsule or sporangium, a slender stalk and a foot (not visible) imbedded in the top of a leafy gametophyte. The capsule is often covered by a cap or calyptra which is an enlarged remnant of the archegonium. Identify capsule, stalk, and calyptra.

Based on the alternation of generation life cycle, what will be produced within the sporophyte's

capsule? _____ By what process will the haploid spores be produced?

_____ Remove a fresh sporophyte from its attachment to the gametophyte and observe it with a stereoscopic microscope. A green stalk indicates a young sporophyte and a brown stalk indicates a mature sporophyte. Carefully remove the cap which is a remnant of the

gametophyte. Are the cells of the cap haploid or diploid? _____ Remove the capsule and transfer it to a drop of water on a clean slide. Open the capsule, place a coverslip on the preparation and observe it with the compound microscope. The small cells are either thick coated spores or cells that will eventually become spores after meiosis occurs.

Spores represent a means of dispersal for bryophytes. The thick coat provides protection from harsh environmental conditions such as drying out.

When the spores are in a favorable environment, they germinate and begin to form a filamentous structure called a *protonema*. Examine the prepared slide of a moss protonema w.m. In what way does the appearance of the protonema provide evidence of the evolutionary ancestry of

mosses? _____
Small bud-like structures may be present along the protonema. These will develop into the "leafy" gametophytes.

IV. Vascular Plants

Several major evolutionary trends can be seen in the vascular plants. The development of vascular tissue allows for increased efficiency in the conduction of water and organic material throughout the growing plant body. Vascular tissue also provides support allowing the plant body to grow taller. Another major trend in the vascular plants is the emerging dominance of the sporophyte generation in both size and complexity. The gametophyte generation, on the other hand, is reduced dramatically. In fact in the gymnosperms and angiosperms the gametophyte consists of only a few cells and develops within the body of the sporophyte.

Vascular plants without seeds

A. Whisk Ferns (Psilophyta) Fig. 10–4

Psilotum is one of two genera of "whisk ferns" that occur in the subtropics and tropics. The sporophyte attains a height of 8 to 36 inches. There is a rhizome which supports upright dichotomous branches. There are no roots present, only rhizoids. The erect stems are photosynthetic and bear sporangia. There are no leaves present, only scales. The gametophyte plant grows underground. It is about ½ inch long. It probably obtains food through a mutualistic association with fungi. Numerous archegonia and antheridia are found on the surface. Water is required for fertilization. After fertilization an embryo develops and grows into a new sporophyte plant.

If available examine a live or preserved specimen.

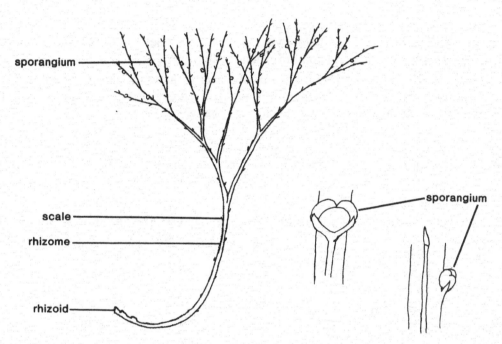

Figure 10–4. *Psilotum* (whisk fern).

B. Club Mosses (Lycophyta) Fig. 10–5

Lycopodium is a representative of a group known as the "club mosses". Representatives of this genus are found in temperate climates, but attain their greatest diversity in the tropics. The sporophytes have above ground stems, runners, roots and spirally arranged leaves.

Sporangia form in the axils of specialized leaves called *sporophylls*. In some species the sporophylls are localized in a distinct cone or *strobilus*, in others they are not. An inconspicuous gametophyte alternates with the sporophyte in the life cycle.

Observe specimens with and without a strobilus. Locate roots, runners, aerial stems, leaves and sporangia.

C. Horsetails (Sphenophyta) Fig. 10–6

Commonly known as "horsetails" and "scouring rushes", members of this genus are both temperate and tropical in distribution. The sporophyte has rhizomes from which aerial branches originate. Aerial branches rarely exceed 4 feet in height in temperate climates. Heights exceeding 15 feet have been recorded in the tropics. The stems are jointed and impregnated with silica. The whorled leaves are scale-like. Branches arise from the node in between leaves and

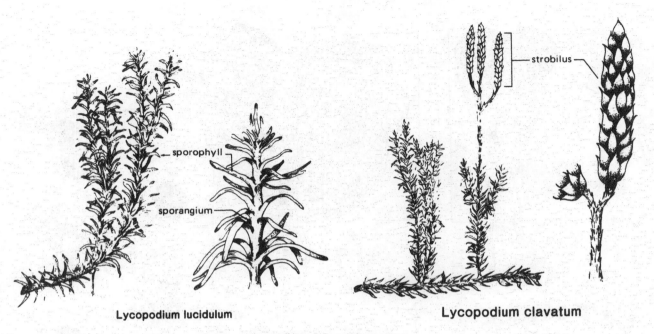

Figure 10–5. Club mosses.

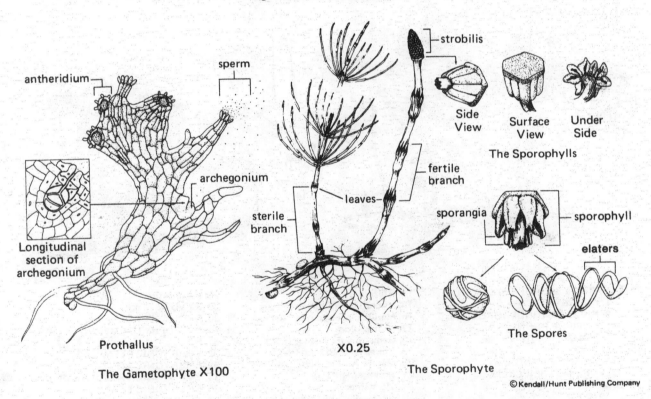

Figure 10–6. *Equisetum arvense* (horsetail).

thus alternate with them. True roots are present. Sporangia are produced on umbrella shaped sporophylls in a strobilus located at the end of an aerial stem. The gametophyte contains chlorophyll and is photosynthetic. It is only a few millimeters in diameter. Motile sperm produced in antheridia require the presence of water to swim to an egg in an archegonium.

Observe a mature sporophyte and locate rhizomes, roots, aerial stems, leaves and strobili.

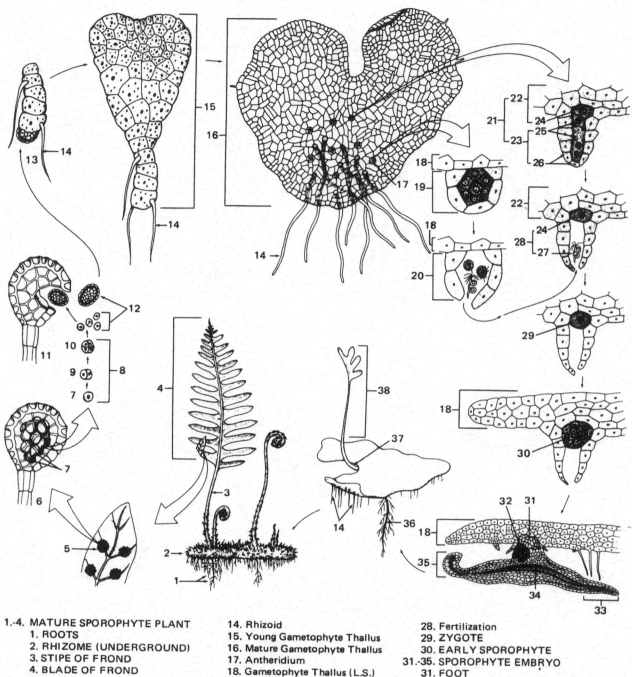

1.-4. MATURE SPOROPHYTE PLANT
  1. ROOTS
  2. RHIZOME (UNDERGROUND)
  3. STIPE OF FROND
  4. BLADE OF FROND
5. SORUS
6. YOUNG SPORANGIUM
7. SPORE MOTHER CELLS
8. *Meiosis*
  9. Dyad
  10. Tetrad of Meiospores
11. MATURE SPORANGIUM
12. Meiospores
13. Germinating Meiospore

14. Rhizoid
15. Young Gametophyte Thallus
16. Mature Gametophyte Thallus
17. Antheridium
18. Gametophyte Thallus (L.S.)
19. Young Antheridium (L.S.)
20. Mature Antheridium (L.S.)
21. Archegonium (L.S.)
  22. Venter
  23. Neck
  24. Egg Cell
  25. Neck Canal Cells
  26. Neck Wall Cells
  27. Mature Sperm

28. Fertilization
29. ZYGOTE
30. EARLY SPOROPHYTE
31.-35. SPOROPHYTE EMBRYO
  31. FOOT
  32. EMBRYONIC STEM
  33. EMBRYONIC ROOT
  34. EMBRYONIC VASCULAR STRAND
  35. EMBRYONIC LEAF
36.-38. YOUNG SPOROPHYTE
  36. ROOT
  37. RHIZOME
  38. LEAF = FROND

Figure 10–7. Life cycle of the fern.

D. Ferns (Pterophyta) Fig. 10–7

Over 9,000 species of ferns are distributed worldwide with the greatest diversity occurring in the moist tropics.

Fern Sporophyte—The fern sporophyte typically consists of an underground stem called a rhizome to which are attached large above ground leaves (fronds). The leaves are coiled and unfold in a characteristic manner. In the coiled state they are referred to as "fiddleheads". Meiosis and spore production take place on the underside of leaves in sporangia. The sporangia are borne in clusters called sori (singular sorus) which appear as discrete dots over the leaf surface or at the margin of the leaf. Leaves that bear sporangia are called sporophylls.

Examine a fern sporophyte and locate the rhizome, leaves and sori. Note the presence of veins on the frond. These represent bundles of vascular tissue. What advantages does vascular tissue

give the plant? _____

Do all of the fronds have sori on their undersurfaces? _____ If several fern species are available, you may notice that some of them have sori on each frond while others do not. The latter trend continues in the seed plants in which certain leaves have evolved into highly specialized sporophylls (cones or flower parts) devoted exclusively to reproduction.

Carefully remove a sorus and place it in a drop of water on a microscopic slide. Add a coverslip and observe it under low magnification. Many sporangia should be present. Note that some of the cells of the sporangial case have cell walls with additional thickenings. These specialized cells play a role in the expulsion of the spores when they lose water. The cells shrink, then bend back and flip forward in a catapult-like action. Do any of the sporangia appear to have released their spores? Make a sketch of a closed sporangium and an opened sporangium if present.

Next, examine a prepared slide labeled fern leaflet mature sporangia or *Pteridium* leaflet with mature sporangia. Identify the vascular tissue within the leaflet (use Fig. 10–8 as a guide). On the underside of the leaflet identify a sorus with sporangia.

Fern Gametophyte—A meiospore on germination develops into a small, heart-shaped gametophyte (=prothallus). The gametophyte lacks vascular tissue, is green and carries on photosynthesis. On the under surface are numerous rhizoids, which anchor the gametophyte to the forest floor, and multicellular reproductive organs. Antheridia are scattered among the rhizoids. Ar-

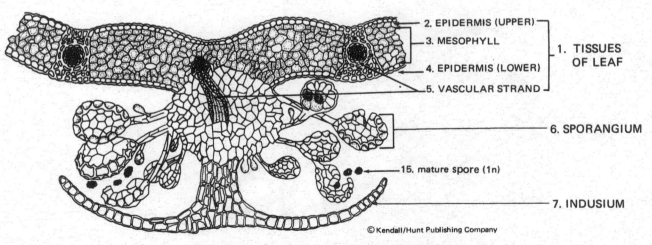

2. EPIDERMIS (UPPER)
3. MESOPHYLL
4. EPIDERMIS (LOWER)
5. VASCULAR STRAND
1. TISSUES OF LEAF
6. SPORANGIUM
15. mature spore (1n)
7. INDUSIUM

Figure 10–8. Cross section of a sporophyll through a sorus.

chegonia, in which eggs are produced, are fewer in number and occur near the notch of the heart-shaped gametophyte. Water is necessary for fertilization as the sperm must swim to the egg. The fertilized egg or zygote develops into an embryo, which soon grows out of the archegonium, takes root and grows into a mature sporophyte.

Obtain and examine a prepared slide labelled fern prothallium, antheridia and archegonia. Draw a prothallium and label rhizoids, antheridia and archegonia. Examine a fresh specimen if available.

Many of the lower vascular plants are homosporous—form one type of meiospore and consequently one type of gametophyte bearing both male and female reproductive structures. Motile sperm are formed and wet conditions are required to accomplish fertilization.

Some lower vascular plants developed two types of spores and consequently two types of gametophytes. This condition, known as *heterospory,* was prerequisite for the development of the seed habit. The female gametophyte is retained within an ovule (sporangium) on the sporophyte. The male gametophyte inside pollen is shed and carried close to the female gametophyte inside an ovule during pollination.

Vascular plants with seeds

The seed plants are subdivided into two groups. The gymnosperms with naked seeds and the angiosperms with enclosed seeds.

Gymnosperms—naked seeds

A. Cycads (Cycadophyta)

Modern cycads occur only in the tropics and subtropics. Nine genera and about 100 species are known. They are palmlike in appearance and are not usually taller than 6 feet. In most genera the pollen and ovules are borne on specialized leaves (sporophylls) in a strobilus. Pollination is accomplished by the wind.

If available, examine a live specimen of **Zamia floridana**.

B. *Ginkgo* (Ginkgophyta) Fig. 10–9

Ginkgo biloba, the maidenhair tree, is the only representative of this group. It is a large tree that is widely cultivated in the United States. *Ginkgo* trees are deciduous and the leaves have

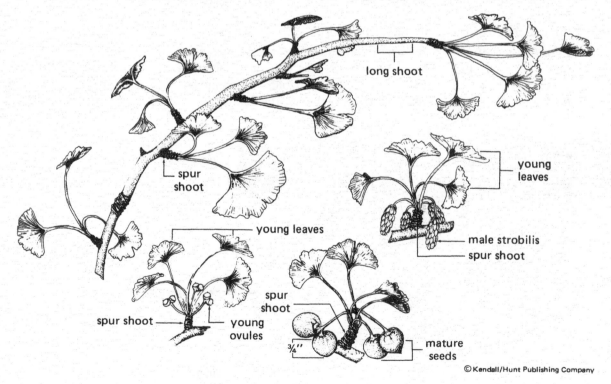

© Kendall/Hunt Publishing Company

Figure 10–9. *Ginkgo biloba.*

dichotomous venation. Microsporangia and ovules occur on separate trees. Pollen is produced in paired microsporangia on sporophylls in a strobilus. Ovules are formed in pairs at the end of spur shoots. Pollination is accomplished by the wind.

Examine specimens of Ginkgo and locate leaves, paired ovules and a male strobilus.

C. Conifers (Coniferophyta) Fig. 10–10

This group is represented by about 550 species of large trees and shrublike plants including pines, fir, spruce, redwood, cedar, hemlock, larch, juniper and yew.

The pine will be used to represent conifers (Figure 10–10). The pine tree or sporophyte is the conspicuous part of the life cycle. Male and female gametophytes are produced. The female gametophyte is dependent on the sporophyte, and, therefore, does not develop independently as in ferns. The male gametophyte (pollen) is shed and carried by the wind. Pollination brings the pollen close to the egg. Nonmotile sperm are produced inside the pollen and, therefore, a film of water is not necessary for fertilization as is the case in bryophytes and lower vascular plants. The pine tree (sporophyte) is differentiated into roots, stem and leaves (needles). Two types of spores are produced on two types of cones: male or staminate (pollen) cones and female or ovulate (seed) cones.

1. Staminate cones—staminate cones appear in the spring, and are borne in clusters. Each staminate cone consists of a central axis to which are attached numerous sporophylls. Two sporangia are formed on the underside of the sporophylls. Meiosis and pollen production take place in the sporangia. Examine a pine branch bearing staminate cones and note their clustered arrangement. Next, examine a prepared slide of a staminate cone labeled *Pinus, l.s. staminate cone with mature pollen grains*. Identify the central axis bearing sporophylls, the sporophylls, the sporangia containing pollen and the pollen. (The male gametophyte inside pollen consists of four cells at the time it is shed. Prior to fertilization one of those cells, the generative cell, divides to form two nonmotile sperm cells.) Make a sketch of a pollen grain.

2. Ovulate cones—ovulate cones are the obvious cones used for decorations at Christmas time. New ovulate cones are produced in the spring. Two ovules (sporangia) are borne on the upper side of scales attached to a central axis. The female gametophyte is produced inside the ovule and is dependent upon the sporophyte. The gametophyte produces two to five archegonia inside of which the eggs are formed. Pollination brings the sperm containing pollen close to the female gametophyte. Fertilization normally occurs about a year later with the seeds maturing and being shed at the end of two years. The seeds (derived from the ovule and containing an embryo) are not enclosed. Examine ovulate cones of different ages.

Angiosperms—seeds enclosed within a fruit (Figure 10–11)

D. Flowering Plants (Magnoliophyta)

As you have seen in the gymnosperms, the sporophyte generation has evolved into a very complex structure and is the dominant phase in the life cycle. Certain leaves (sporophylls) of the plant have evolved into highly specialized structures (stobili, cones) on which the sporangia are located. The sporophyte generation produces two different types of spores (megaspores and microspores). The spores are not released as they were in the bryophytes and seedless plants, but are retained within the sporangium. Within the sporangium the spores produce the gametophyte generation. The megaspores produce the female gametophyte and the microspores produce the male gametophytes. The male gametophyte is contained within the pollen grain. Pollen is re-

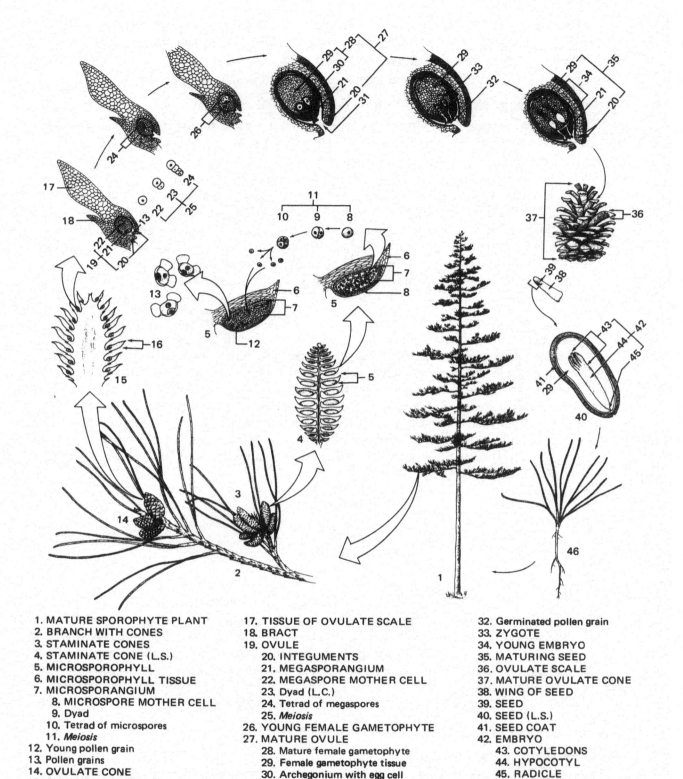

1. MATURE SPOROPHYTE PLANT
2. BRANCH WITH CONES
3. STAMINATE CONES
4. STAMINATE CONE (L.S.)
5. MICROSPOROPHYLL
6. MICROSPOROPHYLL TISSUE
7. MICROSPORANGIUM
   8. MICROSPORE MOTHER CELL
   9. Dyad
  10. Tetrad of microspores
  11. *Meiosis*
12. Young pollen grain
13. Pollen grains
14. OVULATE CONE
15. OVULATE CONE (L.S.)
16. OVULATE SCALE

17. TISSUE OF OVULATE SCALE
18. BRACT
19. OVULE
  20. INTEGUMENTS
  21. MEGASPORANGIUM
  22. MEGASPORE MOTHER CELL
  23. Dyad (L.C.)
  24. Tetrad of megaspores
  25. *Meiosis*
26. YOUNG FEMALE GAMETOPHYTE
27. MATURE OVULE
  28. Mature female gametophyte
  29. Female gametophyte tissue
  30. Archegonium with egg cell
  31. Micropyle

32. Germinated pollen grain
33. ZYGOTE
34. YOUNG EMBRYO
35. MATURING SEED
36. OVULATE SCALE
37. MATURE OVULATE CONE
38. WING OF SEED
39. SEED
40. SEED (L.S.)
41. SEED COAT
42. EMBRYO
  43. COTYLEDONS
  44. HYPOCOTYL
  45. RADICLE
46. YOUNG SPOROPHYTE

Figure 10-10. Life cycle of pine.

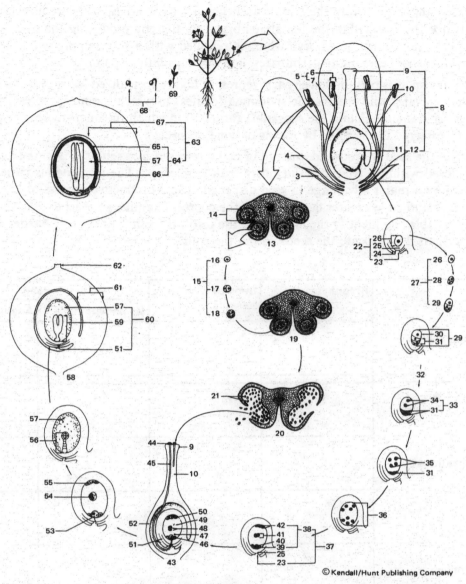

1. MATURE SPOROPHYTE
2. FLOWER
   3. SEPAL
   4. PETAL
   5. STAMEN
      6. ANTHER
      7. FILAMENT
   8. PISTIL
      9. STIGMA
      10. STYLE
      11. OVULE
      12. OVARY
13. YOUNG ANTHER (C.S.)
14. MICROSPORANGIUM
15. *Meiosis*
   16. MICROSPORE MOTHER CELL
   17. Dyad
   18. Tetrad of Microspores
19. MATURING ANTHER (C.S.)
20. MATURE ANTHER (C.S.)
21. Pollen grains (male gametophyte)
22. YOUNG OVULE (L.S.)
   23. MICROPYLE
   24. INTEGUMENTS

25. MEGASPORANGIUM
26. MEGASPORE MOTHER CELL
27. *Meiosis*
28. Dyad
29. Linear tetrad of megaspores
30. Functional megaspore
31. Disintegrating megaspores
32. *Mitosis* (Functional megaspore)
33. Young embryo sac (female gametophyte)
34. Haploid nuclei
35. 4-Nucleate embryo sac
36. 8-Nucleate embryo sac
37. Mature ovule
38. Mature embryo sac
39. Synergids
40. Egg cell
41. Polar nuclei
42. Antipodal nuclei
43. *Double fertilization*
44. Germinated pollen grain
45. Pollen tube
46. Egg cell
47. Sperm nucleus
48. Polar nuclei

49. Sperm nucleus
50. Antipodals
51. INTEGUMENTS OF OVULE
52. OVARY WALL
53. ZYGOTE *(After Fertilization)*
54. ENDOSPERM NUCLEUS (3N)
55. Deteriorating antipodals
56. YOUNG SPOROPHYTE EMBRYO
57. ENDOSPERM TISSUE (3N)
58. YOUNG FRUIT
59. EMBRYO DIFFERENTIATED
60. YOUNG SEED
61. WALL OF FRUIT
62. REMNANTS OF STYLE
63. MATURE FRUIT
64. SEED
65. SEED COAT
66. EMBRYO
67. PERICARP (FRUIT WALL)
68. SEED GERMINATING
69. YOUNG SPOROPHYTE

Figure 10–11. Life cycle of an angiosperm.

leased and is carried to the female archegonium by wind. Sperm (produced within the pollen) fertilize the egg within the female embryo sac, thereby producing the zygote. The zygote, in turn develops into the embryo which is retained with a food supply within a seed. The seed, which is shed, eventually germinates, and develops into a sporophyte.

In angiosperms certain flower parts represent highly specialized leaves devoted to reproduction. The central pistil contains the sporangia (ovules) in which the megaspores are produced. The slender stamens contain the sporangia (anthers) in which the microspores are produced. Pollen is transferred to the pistil by means of wind or insects.

When the seed is produced, instead of being released by itself as it is in the gymnosperms, it is enclosed within a fruit and the entire fruit is released. The tremendous variety of fruit types provides a major advantage in seed dispersal. The reproductive process of the flowering plants will be studied in greater detail later in the course. Examine a model of a flower or fresh flowers if available and identify the anther and the pistil. If fresh fruit is available determine the means of dispersal and locate the seeds within the fruit.

| | | Bryophyte | Seedless Vascular | Seed | |
				Gymnosperm	Angiosperm
	Specimen Name:				
GAMETOPHYTE	body				
	specializations				
	antheridium				
	archegonium				
	method of sperm transfer				
SPOROPHYTE	body				
	specializations				
	sporangium				
	method of dispersal				

REVIEW QUESTIONS

1. Describe four characteristics green algae and land plants have in common.

   a.

   b.

   c.

   d.

2. Diagram and label the basic alternation of generations life cycle.

3. For each plant listed below indicate whether it is a bryophyte, seedless vascular or seed plant.

   a. pine _____

   b. liverwort _____

   c. fern _____

   d. moss _____

4. Name the structure described by each statement that follows:

   a. single haploid cell with thick wall which gives rise to a gametophyte— _____

   b. multicellular structure in which a single egg is produced— _____

   c. multicellular structure in which many sperm are produced— _____

   d. multicellular structure in which spores are produced— _____

   e. multicellular diploid plant body produced from the zygote— _____

5. Describe at least three ways in which bryophytes are adapted to life on land.

   a.

   b.

   c.

6. Describe one major reproductive requirement of bryophytes that is a limitation for land life.
   _____

7. What is the evolutionary significance of the appearance of the moss protonema? _____
   _____

8. State the function of each of the following:

   a. cuticle

   b. stomata

   c. rhizoids

   d. vascular tissue

9. The fern prothallus represents what phase in the alternation of generations? _____ .
Name three structures that may be found on the undersurface of the prothallus.

    a.

    b.

    c.

10. What is a sorus and where can it be found? _____

11. For each plant below indicate whether the gametophyte or sporophyte represents the dominant phase in the life cycle:

    a. liverwort

    b. moss

    c. fern

    d. pine

12. For each plant below name and describe the location of the specific structures in which meiosis occurs. (Hint: Recall from the life cycle what cells are produced from meiosis)

    a. moss

    b. fern

    c. pine

13. Name the multicellular structure found within pollen. _____ What reproductive advantage does a plant have that is able to produce pollen?

EVOLUTION OF ANIMALS I

Objectives

1. Trace the development of the following evolutionary trends and adaptations in each phylum studied:
   a. increased complexity of body plan
   b. body symmetry leading to bilateral dominance
   c. body cavity development
   d. cephalization
2. Identify representative organisms from each phylum studied from slides, models, diagrams, and living and preserved specimens.
3. Provide evidence which supports the evolutionary relationships and origins of the phyla studied.
4. Describe the following adaptations and state their evolutionary significance:
   a. nematocysts
   b. ciliated bilateral larvae
   c. cuticle
   d. specialized appendages
5. Identify the set of characteristics that distinguishes each of the phyla studied.
6. Answer all review questions.

I. Introduction

Invertebrates are animals which do not possess a vertebral column. Approximately 90 percent of all of the described species of animals are regarded as invertebrates. Evolution of the invertebrates is believed to have begun with an ancestral unicellular form that reproduced asexually and sexually. Sexual reproduction produced offspring that differed slightly from each other. Due to competition for food and space and due to predation, many offspring did not survive long enough to reproduce. Only those individuals that possessed "favorable" characteristics for their particular environment were able to survive and pass their genes on to the next generation.

"Favorable" characteristics (called adaptations) in one environment might be "unfavorable" in a different environment. Within each environment a natural selection process determined which of the competing species survived. As this process continued over hundreds of millions of years, many species became extinct, many new species evolved, and some species, having adapted well to a relatively constant environment, remained relatively unchanged.

During this laboratory session you will observe some of the results of the evolution of invertebrates by examining representatives of some of the major invertebrate phyla.

II. Evolutionary Trends and Special Adaptations

A few of the major trends and special adaptations are below. Keep these in mind as you observe the laboratory specimens.

A. Evolutionary Trends
   1. *body construction:* unicellular, multicellular with groups of cells organized in specialized tissue layers, tissues organized as organs, organs grouped into organ systems

2. *body symmetry:* asymmetrical, radially symmetrical, bilaterally symmetrical

3. *body cavity* (coelom): no actual body cavity other than some type of gut (acoelomate), body cavity present, but not completely lined with mesoderm tissue (pseudocoelom), body cavity completely lined with mesoderm (true coelom)

4. *segmentation:* repetition of body parts with increased specialization of segments

5. *cephalization:* specialized sensing organs clustered at anterior end

III. Phylum Cnidaria (Coelenterata)

A. Body Plan

Organisms in the phylum Cnidaria represent the *tissue level* of organization. A tissue is a group of cells acting together in a coordinated fashion. Cnidarians have three basic tissue layers: an outer layer (*epidermis*), an inner layer (*gastrodermis*), and an area in between (*mesoglea*). The epidermis is a well developed layer of closely packed cells which serves to protect the organism and obtain food. The gastrodermis also is well developed and serves in digesting food and in internal transport. The mesoglea, on the other hand, is poorly developed, containing a few scattered cells and much gelatinous matrix. Cells in the mesoglea serve to coordinate the actions of the organism and to produce the gametes (sex cells).

The cnidarians are *radially symmetrical* because any plane cut through the mouth opening along the longitudinal axis of the body will result in mirror images. In an organism that is stationary or one which is carried by currents, this adaptation allows the organism to sense and react to environmental changes coming from any direction.

The representative cnidarian you will be studying is the *Hydra*. Obtain a *prepared slide* of a hydra cross section and label on Figure 11–1: epidermis, mesoglea, and gastrodermis.

The central cavity on the slide is called the gastrovascular cavity. This cavity serves as the gut of the hydra and also its means of internal transport. Some of the gastrodermal cells lining the

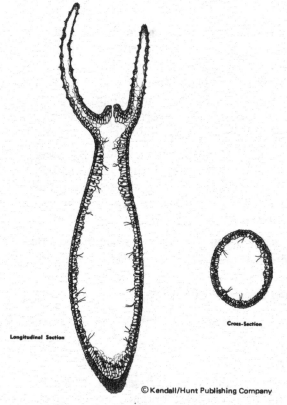

Longitudinal Section

Cross-Section

Figure 11–1. Hydra, sections.

126

gastrovascular cavity have flagella. What is their most probable function? _____

Other gastrodermal cells secrete digestive enzymes into the _____
cavity. This enables the hydra to hydrolyze larger food particles into smaller units which are then engulfed by other gastrodermal cells and digested completely.

B. Adaptations for Reproduction

The hydra reproduces both *asexually* and *sexually*. When environmental conditions are favorable, asexual reproduction takes place. A bud forms, grows and eventually separates to form a new hydra. During unfavorable conditions (i.e., increases in the CO_2 concentration of the water) certain cells in the mesoglea produce either sperm or eggs. Examine the slide labeled *Hydra* budding, and observe the adult and the bud.

C. Adaptations for Feeding and Defense

1. Obtain a live hydra from the stock table and by using a Pasteur pipette, carefully transfer it to a syracuse dish. Note: the hydra should be placed only in filtered, *aerated pond water*. Do not use tap water.

2. Place the syracuse dish on the stage of a dissecting scope and observe the behavior of the hydra. The hydra is a sedentary animal; therefore, it will probably attach itself to the dish.

3. After the hydra attaches, take a clean set of forceps and place a granule of agar (that had been soaking in a .03 gram/liter solution of reduced glutathione) onto the tentacles of the hydra.

4. Record and diagram the actions of the hydra in the space below.

5. If live microcrustaceans are available, such as *Daphnia,* you or your instructor can introduce some into the dish with the *Hydra*.

6. Watch the action of the hydra closely. Certain cells of the ectoderm contain stinging capsules called *nematocysts*. The nematocyst contains a coiled hollow thread. When stimulated, the nematocyst thread is released. Some threads are very long and coil around the prey while others contain a toxin which paralyzes the prey.

In addition to releasing the nematocyst threads, the hydra may have also changed its elongated shape into a ball. This coordinated effect was produced through the action of a *nerve net*. Special cells of the ectoderm are able to sense changes in the environment. These *sensory cells* in turn induce an electrical impulse in nerve cells located in the mesoglea. The impulse travels from nerve cell to nerve cell and is finally conveyed to all cells of the body. Each cell is equipped with a *contractile fiber* as its base, which, when stimulated, allows the hydra to change its shape. The action of the nerve net is also important in the movement of the tentacles in feeding and in the occasional summersault movements of the hydra.

D. Variations Among the Cnidarians

Jellyfish, sea anemones, and true coral are also cnidarians.

Stationary forms of cnidarians are called *polyps* and free-swimming forms are called *medusae*. Some cnidarians in their life cycle produce a polyp which produces medusae asexually. The medusae each produce either male or female gametes which are released into the water. Fertilization occurs in the water with the resulting zygote eventually developing into a free-swimming ciliated larva (planula). After swimming for a while, it attaches to a substrate and develops into a polyp form. This illustrates an adaptation of a sessile organism which allows for dispersal of the species. See Figures 11–2, 11–3, and 11–4.

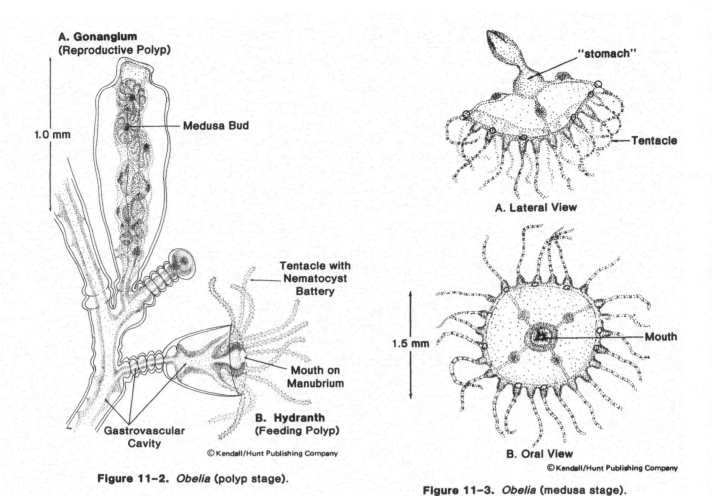

A. Gonangium
(Reproductive Polyp)

1.0 mm

Medusa Bud

Tentacle with
Nematocyst
Battery

Mouth on
Manubrium

Gastrovascular
Cavity

B. Hydranth
(Feeding Polyp)

© Kendall/Hunt Publishing Company

Figure 11–2. *Obelia* (polyp stage).

"stomach"

Tentacle

A. Lateral View

1.5 mm

Mouth

B. Oral View

© Kendall/Hunt Publishing Company

Figure 11–3. *Obelia* (medusa stage).

Tentacle

Mouth

Gastrovascular Cavity

Septum

Lateral View (One Quarter Cut Away)

Cross-Section

© Kendall/Hunt Publishing Company

Figure 11–4. *Metridium*.

E. Evolutionary Possibilities

Examine the slide, *Aurelia* planula w.m. and diagram the planula. Does it possess radial symmetry or bilateral symmetry? Does its bilateral symmetry coupled with its ciliated surface provide it with an adaptive advantage in moving? Could an ancestral planuloid organism give rise to a completely new line of bilateral organisms? Compare the planula with the members of the next phylum.

IV. Phylum Platyhelminthes—the Flatworms

A. Body Plan

Members of the phylum *Platyhelminthes* (flatworms) are on the *organ-system level* of organization. A structure composed of different types of tissues acting together is called an *organ*. An organ-system is one or more organs acting with other structures in the performance of life characteristics. The live specimen you will study from the phylum *Platyhelminthes* will be the free-living *planarian*.

1. Transfer a small amount of water from the stock culture of planaria into a syracuse dish.

2. Next, using a glass rod, transfer a live planarian into the dish.

3. Observe it under a dissecting scope.

4. Diagram the shape of the planarian below:

This type of a body shape is known as *bilateral symmetry,* since there is *only one plane* in which the body can be divided into mirror images. Do you think that the planarian could have evolved from a planuloid ancestor?

Place a slide of a planarian cross section on the microscope. On Figure 11–5 label the gastrovascular cavity and the three body layers: epidermis, mesoderm, and gastroderm.

The middle body layer (mesoderm) has advanced quite a bit from the primitive mesoglea of the hydra. In addition to providing bulk and support for the organism, the mesoderm gives rise to muscle tissue and highly complex reproductive organs.

B. Cephalization

Bilateral symmetry is of great adaptive value in a free-moving organism. The area (anterior end) of the animal which is first to contact new environments as the animal moves can be greatly specialized. The planarian exhibits cephalization, which is the accumulation of sense organs at the anterior end. Identify the two large "*eye spots*." Label these on your diagram. These are actually sense organs which detect changes in light intensity and aid in the orientation of the planarian. Shine a concentrated light on the anterior end of the planarian and note its reaction. The planarian should move away from the light.

Identify two lateral knobs or lobes on each side of the head area. These lobes are extremely sensitive to touch, chemicals, and even water currents. Place the tip of a dropper into the dish

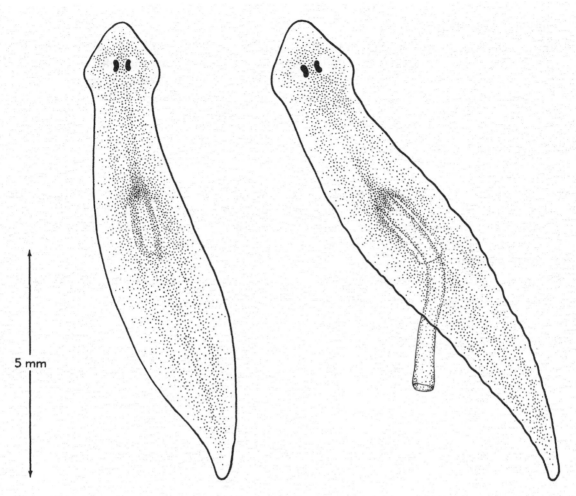

5 mm

External View

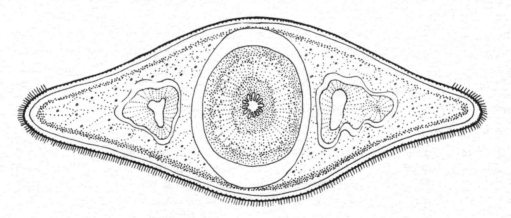

Cross-Section Through Pharynx

Figure 11–5.

containing the planarian and gently create a water current by pressing the rubber tip of the dropper. Apply the current in such a manner that it strikes the planarian at the region where the sensory lobes are located. Note the reaction. _____

The water current striking the sensory lobes stimulates the planarian to orient *toward* the current. Now change the direction of the current, but this time direct it to the middle or posterior part of the body. Does the planarian respond this time? _____

If the current has not struck the sensory lobes the planarian should not change direction. What is the reaction if the current is applied from the rear? _____

_____

The planarian responds to this stimulus since the lobes project laterally from the head and thus can "sense" the current coming from the rear.

C. Adaptations for Food Procurement and Digestion

Place some egg yolk in the dish and try to observe the *feeding behavior* of the planarian. The planarian is a free-living organism and feeds off the living or dead small animals that it sucks up from the bottom of ponds or quiet streams.

The main organ involved in the actual *procurement of food* is a *muscular pharynx* on the ventral surface just behind the mouth. Obtain a prepared slide of a planarian (*whole mount*) and observe it under low power. Try to locate the pharynx. As the planarian crawls over a substrate, the pharynx protrudes from the body and by means of muscle contraction, sucks up the food.

The *food* passes through the *mouth* and *pharynx* and then enters a *gastrovascular cavity*. The gastrovascular cavity of the planarian differs from that of the hydra in that it is *branched*. Since the only means of internal transport in the planarian is the gastrovascular cavity, it is a necessity to have the large amount of branching in order to deliver nutrients to all cells of the body. However, like the hydra, there is only one opening in the gastrovascular cavity. Both food and wastes must pass through this one opening.

D. Variation Among the Platyhelminthes

The planarian is a member of the Class *Turbellaria*. This class includes all flatworms which are free-living. There are two other major classes in the Phylum Platyhelminthes, Class *Trematoda* (flukes) and Class *Cestoda* (tapeworms). Both of these classes are made up of parasitic organisms and will be studied in a later laboratory.

V. Phylum Nematoda—the Unsegmented Roundworms

This phylum consists of wormlike organisms, some of which are parasitic while others are free-living. The major evolutionary advancement in this phylum is the presence of a *complete digestive* system. A complete digestive system is one in which there are two openings, a mouth and an anus. This allows food material to move in a one-way direction thereby eliminating the mixing of predigested incoming food and indigestible waste material.

A second advancement is the presence of a cavity (*pseudocoelom*) between the outer body wall and the digestive tube. This cavity provides a degree of protection for internal organs by acting as a shock absorber or cushion, and allows some degree of movement for internal organs.

Examine the microscope slide *Ascaris* c.s. Begin at the outer cuticle and identify the hypodermal layer below the cuticle which secretes the cuticle, a layer of muscle, the pseudocoelom and the intestinal cavity.

What advantage is a non-living cuticle in a parasitic organism, especially if the infection route is by way of mouth?_____

The muscle layer is of mesodermal origin and the intestinal cells are of endodermal origin. Therefore the pseudocoelom which is the cavity between these two layers is lined by tissue of both mesodermal and endodermal origin.

Located within the pseudocoelom are cross sections of the ovary, oviduct and uterus. The uteri have large lumina and contain eggs, the oviduct(s), if present, has a small lumen and the ovaries have no lumina.

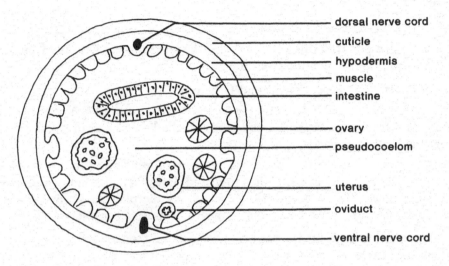

Figure 11-6. Cross section of Female *Ascaris*.

REVIEW QUESTIONS

1. Construct a chart which describes the following items for each phylum studied: body construction, body symmetry, body cavity, segmentation, cephalization. If the trait is completely absent in the phylum place a zero (0) in the appropriate space on the chart.

2. What is the evolutionary advantage of the following adaptations:

   a. nematocysts

   b. ciliated bilateral larvae

   c. cuticle

   d. complete digestive system

3. Give an example of an organism which displays at least one of the adaptations listed in question #2.

4. Identify the phylum in which each of the following organisms is classified:

    a. hydra _____   d. *Ascaris* _____

    b. planarian _____   e. jellyfish _____

    c. tapeworm _____   f. true coral _____

5. Name the phylum or phyla that possess the items below:

    a. radial body symmetry _____

    b. incomplete digestive tract (gastrovascular cavity) _____

    c. nematocysts _____

    d. complete digestive tract _____

    e. non-living cuticle _____

    f. planula larvae _____

    g. pseudocoelom _____

EVOLUTION OF ANIMALS II

Objectives

Upon completion of this exercise you should be able to:

1. State the evolutionary significance of body segmentation and identify evidence of segmentation in each phylum studied.
2. Identify the coelom from microscopic slides or models and state the evolutionary significance of its development.
3. Identify the evidence used to support the evolutionary relationship between annelids and mollusks.
4. Distinguish among the three classes of annelids and recognize examples of each class.
5. Identify four unique characteristics of the phylum Mollusca from specimens, microscopic slides, models and/or drawings.
6. Distinguish among the three major classes of mollusks and recognize examples of each class.
7. Describe the concept of convergent evolution and explain how the squid may be used to illustrate it.
8. Recognize from arthropod specimens, models and drawings, five modifications of the basic body plan of annelids.
9. Identify the principle of adaptive radiation and explain how mollusks and arthropods illustrate it.
10. Identify members of the six classes of arthropods.
11. Answer all questions in the exercise.

I. Introduction

In the previous exercise you traced the development of several evolutionary trends in the animal kingdom such as the increased complexity of the body plan (tissues to organ systems), body symmetry (radial to bilateral) and body cavity development (acoelomic to pseudocoelomic). In this exercise you will continue to follow these trends as well as exploring new evolutionary developments such as segmentation and the development of specialized appendages. Three major phyla will be examined in this exercise, Annelida, Mollusca and Arthropoda. As you perform the exercise, you will see why the arthropods and mollusks are considered to be the two most successful phyla in the animal kingdom.

II. Phylum Annelida—the Segmented Roundworms

A. Body Plan and Adaptions for Movement

Obtain a live earthworm from the demonstration table and place it on a moist paper towel in a dissecting tray. What type of symmetry does it exhibit? _____ Can you distinguish one end from the other? What is the difference? _____

Note the ringed or segmented appearance of the body. As you study the earthworm, note the many adaptive advantages obtained from segmentation. Observe how the worm moves.

Notice that some segments are elongating and some are shortening. As a segment elongates, does it become thinner (constricted) or thicker? When a segment shortens, it also bulges out. Notice how this "bulge" appears to move in a rhythmical coordinated fashion toward the posterior of the worm.

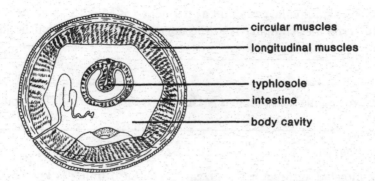

Figure 12-1. Cross section of an earthworm.

This type of locomotion allows the earthworm to travel rapidly through the soil. The evolution of a segmented body is largely responsible for this specialized movement. Examine the slide showing a cross section of the earthworm or observe the model on the demonstration table. Identify the following components of the body wall: epidermis, circular muscle layer, and longitudinal muscle layer.

Contraction of the outer circular muscles elongates the body and contraction of the inner longitudinal muscles shortens the body. Replace the live earthworm and obtain a preserved specimen.

B. Adaptations for Digestion and Absorption

Place a moist paper towel in the dissecting tray and arrange the worm so that the darker dorsal surface is facing up. Place a pin through the first and last segment. Now fill the pan with just enough water to cover the worm. Very carefully using pointed scissors, cut from segment one to about segment 49. Make the cut just to the left of the dark line running down the middle of the dorsal surface. Only cut the body wall! Do not disturb the internal organs! After you have made the cut, pin the lateral flaps to the pan. Identify the following:

1. Septa (transverse partitions which divide the worm into segments).

2. Digestive tract (tube extending the length of the worm).

3. Coelom (space between body wall and digestive tract).

Again observe the slide of the cross section of the earthworm. Locate the coelom and note the presence of the mesodermal-derived lining. View the intestinal wall under high power. The innermost layer is gastrodermal tissue. Just outside the gastoderm are two layers of muscles which are of mesodermal origin. Rhythmic contraction of these muscles (peristalsis) allows food to be moved through the digestive tract without moving the entire animal. This advantage (independence of the digestive tract from the muscles of the body wall) is found only in organisms with a true body cavity (coelom).

Beginning at the mouth, identify the following specialities of the digestive tract:

1. Just behind the mouth the digestive tract is known as the *buccal cavity*.

2. In segments four and five, the tract enlarges and appears quite muscular. This is known as the *pharynx*. The muscular pharynx aids in sucking in soil and food.

3. In segments six to 14, the tract is very thin and is called the *esophagus*.

4. Beginning at segment 10, are three pairs of *calciferous glands*. These glands secrete calcium carbonate which neutralizes any organic acids found in the food and helps eliminate excess calcium.

5. The digestive tract enlarges again at segment 15. Make a small cut into this enlarged area called the *crop* and compare it to the muscular pharynx. What is the function of this thin walled crop? _____

_____

6. The *gizzard* begins at segment 17. How does the *gizzard* differ from the crop?

_____

from the pharynx? _____

Feel the internal surface of the gizzard. The hard lining is known as a cuticle and along with sand grains found there, allows the muscular gizzard to grind the food thoroughly. The crop, on the other hand, serves as a storage area.

7. From the gizzard, the food passes into the long *intestine* where it is chemically digested and absorbed. Cut across the intestine and observe that the dorsal wall dips downward into a fold called the *typhlosole*. The typhlosole aids in increasing the surface area for absorption.

8. Soil and undigested material pass out the *anus*.

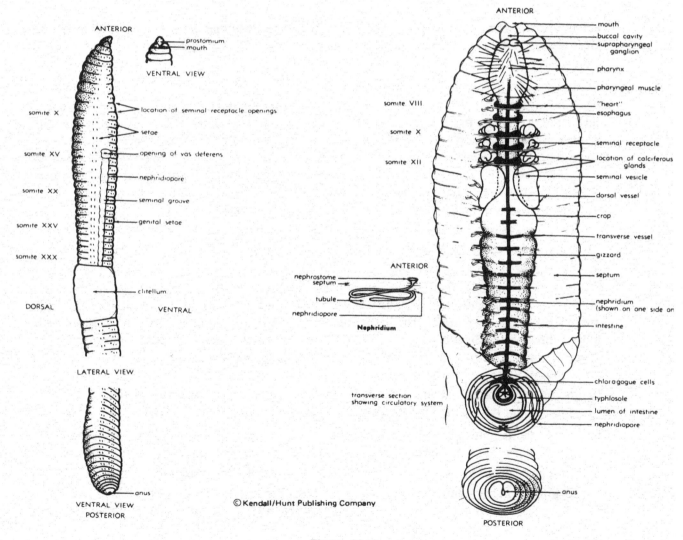

© Kendall/Hunt Publishing Company

Figure 12-2.

C. Variations Among the Annelids

The Phylum Annelida consists of about 9,000 different species divided into three major classes and several minor classes. The earthworm is in the Class Oligochaeta (few bristles). The class, Hirudinea contains the leeches most of which are parasitic. The Class Polychaeta ("many bristles") consists mainly of marine worms. Observe the specimens on demonstration.

D. Evolutionary Relationships

Observe the preserved specimen, *Neanthes,* (Nereis). In which Annelid class does it belong?

_____ On either side of each segment of this marine polychaete is a lobed, flat parapodium ("side foot") which is abundantly supplied with blood vessels. The paired parapodia thus function both in locomotion and gas exchange. Consider the possibilities of natural selection favoring the specialization of different appendages for tasks such as locomotion, protection, sensing, and feeding.

Another evolutionary relationship concerns the annelids and the mollusks. Species in both phyla possess very similar larval forms. The larval form, called trochophore, possesses a characteristic band of cilia around its equator. The cilia provide locomotion as well as bring food to the mouth.

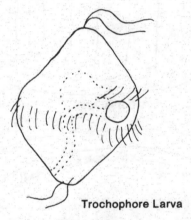

Trochophore Larva

Figure 12–3.

III. **Phylum Mollusca ("Soft Bodied Animals")**

A. Body Plan

Members of the phylum Mollusca have evolved into a highly diverse group of organisms. In general, the phylum exhibits bilateral symmetry and a true coelom that has been substantially reduced.

Although mollusks do not exhibit segmentation, fossil evidence supports ancestral mollusk segmentation. Additional support came in 1952, when living specimens of a clearly segmented mollusk, *Neopilina,* were dredged up off the coast of Costa Rica.

B. Specialization of Mollusks

Mollusks have evolved several distinctive characteristics such as:

1. the development of a ventral muscular foot
2. a fleshy mantle which in most cases secretes a shell
3. a soft dorsal mass containing the internal organs
4. a toothed radula in the mouth which is used to rasp food

Obtain a chiton specimen and identify the calcareous plates which make up the shell. How many

plates are present? _____ What structure secreted the eight plates? _____

Now examine the ventral surface. The broad flat structure is the muscular foot. In life it is covered with a slimy secretion. Note the heavy fold of tissue surrounding the foot. This is the mantle which secretes the plates. Suspended between the mantle and the foot are rows of gills which are used for gas exchange. Move or cut the foot in order to see the gills. The remaining organs of the chiton are embedded within the soft visceral mass just dorsal to the foot and covered by the mantle.

Locate the mouth opening at the anterior end of the ventral surface. Make a deep median cut into the mouth using a sharp razor blade or scalpel. Spread the sides of the mouth apart and

attempt to locate the toothed structure called the r _____ . The radula is attached by muscles and may be extended from the mouth in order to scrape off small fragments of algae.

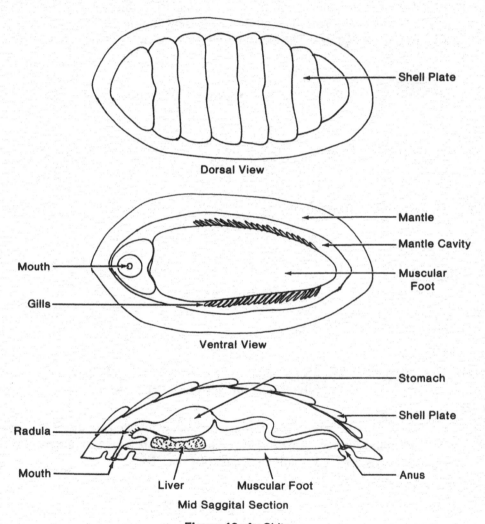

Figure 12-4. Chiton.

C. Variations Among the Mollusks

The Phylum Mollusca consists of about 50,000 species divided into six classes. They have evolved into a diversity of forms which have adapted successfully to salt water, fresh water, and land. The chiton belongs in the Class Polyplacophora ("many plates"). Chitons crawl over and attach to rocks or coral and feed upon algae. What specialization do they possess to hold onto the rock?

_____

What structure do they use to scrape the algae? _____

The largest class of mollusks is Gastropoda which contains snails, slugs, limpets, nudibranchs, etc. Gastropods usually have a single, coiled shell although in some species the shell is absent. The radula is found in the mouth cavity. It is armed with hard teeth set on a long ribbon of tissue which is drawn back and forth to scrape food. Examine a microscope slide of a snail radula. Sketch a portion of the radula.

The majority of the gastropods exhibit a dextral coiling pattern, that is, the shell coils to the right. Only a few species have a shell that is sinistral or coiled to the left. To determine if a snail is "right-handed" or "left-handed" hold the pointed spire up. If the shell opening, or aperture, is to the right of the midline then the coiling is dextral. An opening to the left indicates sinistral coiling. Observe the snails on demonstration and determine their coiling pattern.

Gastropods play an enormous role in the transmission of diseases to humans, domestic and wild animals. Virtually all parasitic trematodes have a molluscan intermediate host in their life cycle. More than 300 million human blood fluke, liver fluke and lung fluke infections occur which could be eliminated if the snail intermediate hosts could be controlled.

Another major class is Bivalvia (or Pelecypoda) which contains the clam, mussels, etc.

Bivalves have two calcareous shells equipped with a hinge to hold them together. Strong adductor muscles can close the shell tightly. All bivalves are aquatic in salt or fresh water. The foot is reduced and laterally compressed. Locomotion is limited and many forms are completely sessile. The soft body mass houses the digestive and reproductive organs. The mantle covers the body mass and secretes the shell. No radula is present in this class. The ciliated gills are used for gas exchange and to filter food out of the water. Observe the clam specimen and model and identify shells, mantle, muscular foot, body mass, and gills. Refer to Figure 12–5.

The third major class of mollusks is Cephalopoda which contains the squid, octopus, nautilus, etc. The cephalopods have the foot greatly modified into tentacles. The shell is either single or absent. The body is streamlined for rapid locomotion. One of the evolutionary trends in this class has been the reduction of the shell. The chambered nautilus has a large, external, calcareous shell. It is chambered because as the animal grows, new chambers are added to the spiral; the animal occupies only the outermost one. Nautiloids were the dominant cephalopods in the ocean about 400 million years ago. Today only two species remain and are found at great depths in the Pacific Ocean. The shell of the squid is reduced, internal and not calcareous. At the end of this evolutionary line is the octopus with no shell present at all. Ancestors of the latter two do not appear in the fossil record until about 150 million years ago.

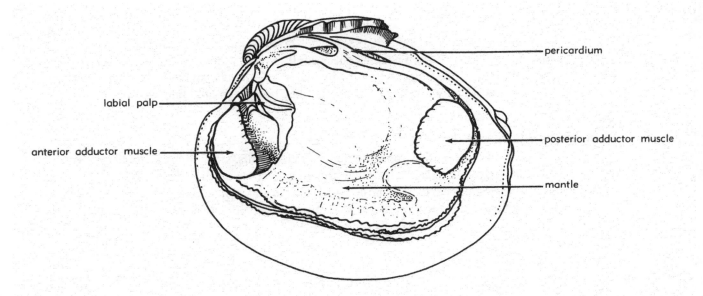

labial palp

anterior adductor muscle

pericardium

posterior adductor muscle

mantle

Left Valve Removed, Mantle Intact

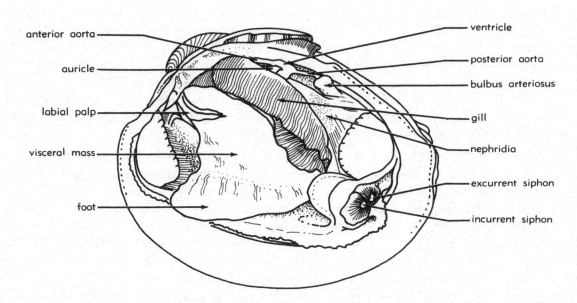

anterior aorta

auricle

labial palp

visceral mass

foot

ventricle

posterior aorta

bulbus arteriosus

gill

nephridia

excurrent siphon

incurrent siphon

Left Valve Removed, Mantle Removed

Figure 12-5. Clam, anatomy.

Observe the squid on demonstration. Note the highly developed eye. Experiments have demonstrated that cephalopods are capable of image formation and shape discrimination. Even though the eye of the squid possesses a cornea, lens, iris and retina, its development is very different from the vertebrate eye formation. The fact that cephalopod and vertebrate eyes independently evolved similar structure and function through different developmental patterns is known as convergent evolution.

IV. Phylum Arthropoda ("Jointed Appendages")

A. Body Plan

The basic body plan of arthropods is derived from annelidlike ancestors in that the arthropods possess segmentation, bilateral symmetry, a complete digestive system, and a true coelom. During the evolution of arthropods natural selection favored several modifications of the annelid plan such as:

1. reduction in the number of segments
2. increased specialization of segments with groups of segments fusing together to form definite body regions
3. modification of the cuticle body covering into a water impermeable exoskeleton composed of chitin
4. reduction of the coelom and presence of a blood filled cavity called the hemocoel
5. presence of highly specialized jointed appendages

B. Specialized Appendages of the Crayfish (Fig. 12–6)

Examine the preserved crayfish specimen and identify the exoskeleton. Note the fusion of the anterior segments to form a cephalothorax (fused head and thorax) region. The remaining body segments are fused to form the abdominal region.

Place the specimen on its dorsal surface (back) and observe the specialized appendages. The specialized appendages are well adapted for sensing the environment, handling, crushing, and tasting food, fighting, walking and swimming, and even reproduction. Identify the paired appendages on the specimen beginning at the posterior end. Remove the appendages from one side of the specimen as you identify them.

First locate the ventral anal opening on the last abdominal segment. The posterior fan-like extension of this segment is called the telson. Lateral to the telson is the appendage called the uropod. It consists of three portions, one which is attached directly to the body, and two which are broad and flat. The uropods and the telson are well adapted for swimming, especially backward swimming.

The next five pairs of appendages are called swimmerets and primarily function in swimming. Before removing the swimmerets, examine all five pairs. Are the first two pairs (i.e., most anterior) different from the rest? If they are elongated and oriented in an anterior direction, then the specimen is a male. These specialized swimmerets are used to transfer sperm to the female. As you remove the swimmerets note that they are composed of a basal portion and two branches.

The next four pairs of appendages are called walking legs. Note that only one branch is present on each walking leg and that each leg attaches to a feathery gill.

The next pair of appendages are called chelipeds and are well adapted for grasping food and fighting.

The next three pairs of appendages are called maxillipeds and are used mainly for handling food. Which maxillipeds have gills attached? _____

The next two pairs of appendages are called maxillae and aid in passing food to the mouth. The larger maxilla is called a gill bailer because it aids the flow of water out from the gill area.

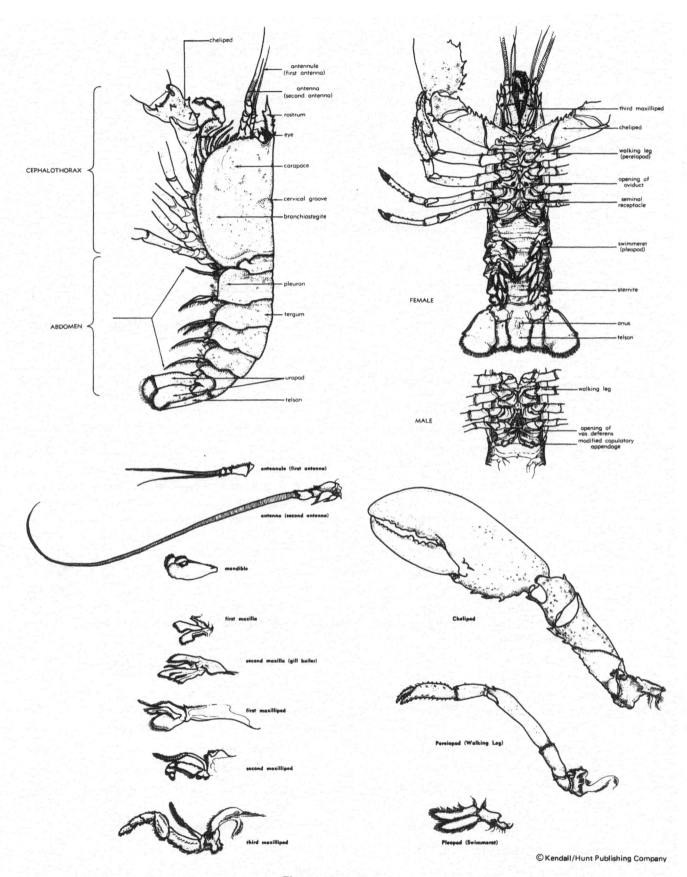

CEPHALOTHORAX

ABDOMEN

cheliped

antennule (first antenna)

antenna (second antenna)

rostrum

eye

carapace

cervical groove

branchiostegite

pleuron

tergum

uropod

telson

third maxilliped

cheliped

walking leg (pereiopod)

opening of oviduct

seminal receptacle

swimmeret (pleopod)

sternite

anus

telson

FEMALE

walking leg

opening of vas deferens
modified copulatory appendage

MALE

antennule (first antenna)

antenna (second antenna)

mandible

first maxilla

second maxilla (gill bailer)

first maxilliped

second maxilliped

third maxilliped

Cheliped

Pereiopod (Walking Leg)

Pleopod (Swimmeret)

© Kendall/Hunt Publishing Company

Figure 12-6. Lobster.

142

The next pair of smooth, shiny appendages are the mandibles and serve to crush food.

Anterior to the mandibles are two pairs of sensory antennae. The longer single pair are called the 2nd antennae and the shorter pair consisting of two filamentous branches are called the 1st antennae.

C. Variations Among the Arthropods

Phylum Arthropoda is the largest phylum (over 1,000,000 species have been classified). Arthropods have undergone adaptive radiation to exist in the ocean, in fresh water, on land, and in the air. The six major classes of Arthropods are: Merostomata (horseshoe crabs), Arachnida (spiders, ticks, mites, etc), Crustacea (crabs, crayfish, lobsters, shrimp, barnacles), Chilopoda (centipedes), Diplopoda (millipedes), and Insecta (insects).

On the demonstration table are representatives of some or all of the six major classes of arthropods. Become familiar with the specimens so that you are able to identify each to a class. Use the following dichotomous key to identify those with which you are not familiar:

1. a. Antennae present 3
    b. Antennae absent 2
2. a. Ten walking legs; tail spike-like .... Class Merostomata
    b. Eight walking legs; no spike-like tail .... Class Arachnida
3. a. Two pairs of antennae .... Class Crustacea
    b. One pair of antennae 4
4. a. Adults with six legs .... Class Insecta
    b. Adults with more than six legs 5
5. a. One pair of legs per body segment .... Class Chilopoda
    b. Two pairs of legs per body segment .... Class Diplopoda

Class Merostomata

The horseshoe crab, *Limulus,* is a representative of this primitive class. Horseshoe crabs are bottom feeders in shallow ocean waters. They have changed little from the appearance of the group in fossil remains 200 million years old to today. The long tail is not used for defense, but rather to aid in righting the animal if it turns over.

Class Arachnida

This diverse terrestrial group includes all spiders, scorpions, harvestmen, ticks and mites. All have eight legs as adults, no antennae and the head and thorax are fused into a cephalothorax. The mouth parts are specialized for sucking so members of this class are predators that kill their prey and then suck the juices out of them or they suck the juices from plants. Many people fear spiders but in the United States, only the black widow is truly harmful. Ticks are much more harmful in that they spread diseases such as Rocky Mountain spotted fever, Lyme disease, babesiosis, Texas cattle fever, etc., to humans and other animals.

Class Crustacea

Almost all crustaceans are aquatic in marine, brackish and freshwater habitats. The possession of two pairs of antennae and gills for gas exchange characterizes members of this diverse class. Many species have commercial value as food for humans including shrimp, crabs, crayfish and lobsters. On a broader scale, crustaceans play an important role in many food chains where they eat plant matter and act as primary consumers and, in turn, are eaten by secondary consumers.

Class Chilopoda

The terrestrial animals known as centipedes or "100-leggers" constitute the class chilopoda. Centipedes probably never have 100 legs! However, centipedes have many legs with one pair per body segment, except the head. All centipedes move rapidly, are predators and have fangs to deliver poison to subdue and capture prey. Turning over a rotting log almost anywhere in the United States will reveal one or more species of orange centipedes of the genus *Scolopendra*. One rather harmless centipede, *Scutigera*, enters houses where it hunts insects on walls and ceilings at night. Despite their being harmless, most wives object to their presence on the bedroom ceiling!

Class Diplopoda

The slow moving, herbivorous, terrestrial millipedes are familiar members of the class Diplopoda. They are most common in forest biomes such as exist in eastern United States and in the moist tropics. Over 150 species are known from Virginia! Millipedes do not inflict painful bites and can be distinguished from their relatives the centipedes by the two pairs of legs per body segment.

Class Insecta

The insects are among the best known and most successful animals. Insects typically have three body regions: a head, thorax and abdomen. On the thorax are attached two pairs of wings and three pairs of legs. Insects are terrestrial as adults, although some species have aquatic larvae. To exchange gases in a dry terrestrial environment, insects have tracheal tubes or tracheae that branch repeatedly to deliver oxygen to cells deep in their body. This inefficient gas exchange system is believed to be a major reason why insects are small.

Observe the insects on display. From the demonstrations and from personal experience, list five examples each of how insects are beneficial to humans and harmful to humans.

	Beneficial	Harmful
1.	_____	_____
2.	_____	_____
3.	_____	_____
4.	_____	_____
5.	_____	_____

REVIEW QUESTIONS

1. Construct a chart which describes the following items for each phylum studied: body construction, body symmetry, body cavity, segmentation, cephalization. If the trait is completely absent in the phylum place a zero (0) in the appropriate space on the chart.

2. What evidence supports the relationship between annelids and mollusks?

3. What evidence supports the relationship between annelids and arthropods?

4. What is convergent evolution? Give an example studied in this exercise.

5. What is the evolutionary advantage of the following adaptations:

   a. coelom

   b. segmentation

144

c. radula

d. specialized appendages

6. Name the phylum or phyla that possess the items below:

   a. mantle _____

   b. exoskeleton _____

   c. parapodia _____

   d. ventral muscular foot _____

   e. jointed appendages _____

   f. trochophore larvae _____

   g. cephalothorax _____

   h. radula _____

7. What evidence of segmentation is demonstrated by the chiton?
   _____

8. If evolutionary success is measured by the number of species in a phylum, which phylum studied is the most successful? _____

EVOLUTION OF ANIMALS III

Objectives

Upon completion of this exercise you should be able to:

1. Distinguish between a deuterostome and a protostome.
2. List the main features which distinguish the echinoderms from the other phyla of the animal kingdom.
3. Identify the anatomical parts of and describe the function of the water vascular system of a starfish.
4. List the three unique characteristics which all chordates exhibit at some stage in their life cycle.
5. Identify notochord, dorsal hollow nerve cord and pharyngeal gill slits on slides labeled Amphioxus w.m. and Amphioxus pharynx c.s.
6. Outline and support a sequence of evolution of vertebrates.
7. Describe at least two evolutionary trends exhibited by the classes: Agnatha, Chondricthyes and Osteichthyes.
8. List and identify at least four adaptations for life on land exhibited by higher vertebrates.
9. Identify at least four features of bird evolution that have enabled them to fly.
10. Identify a vertebrate as to the class in which it belongs (Agnatha, Chondrichthyes, Osteichthyes, Amphibia, Reptilia, Aves, Mammalia) and describe the major features of each class.
11. Identify from skulls, models and drawings, the four types of teeth found in mammals; indicate their primary use and trace the evolutionary trends in mammalian dentition patterns.
12. Answer all the review questions in the exercise.

I. Introduction

In two previous exercises you studied several animal phyla. In this exercise you will examine two additional phyla, the echinoderms and the chordates. These two phyla are believed to have evolved from a coelomate ancestor, but unrelated to the annelid phylogenetic line. The evidence for this difference comes from the study of the embryological development of the two evolutionary lines.

The echinoderms and chordates are called deuterostomes because the mouth develops from the second opening in the embryo. The remaining animal phyla are called protostomes because the mouth opening develops from the first opening in the embryo. Additional developmental differences include the pattern of cell division (cleavage) of the egg and the manner of mesoderm and coelom formation.

II. Phylum Echinodermata "Spiny Skinned Animals"

This phylum represents an odd group of organisms sharply distinguished from all other groups of the animal kingdom. All the members of the Echinodermata live in the sea. Included in this phylum are sea stars, brittle stars, sea urchins, sea cucumbers and sea lilies. Echinoderms pass through a larval stage which is bilaterally symmetrical and then develop into adults having radial symmetry. They appear to have been derived from an ancestral line having bilateral symmetry—probably a flatworm or a flatwormlike organism. The fossil record indicates that echinoderms were once sessile, much as the modern sea lilies are. It was probably at that time that radial body symmetry evolved in the phylum. Now most are motile but are still saddled with radial symmetry even though bilateral symmetry seems to be better suited for a motile organism. Evolution is usually a one way process.

Unique features of the group include:

1. a water vascular system.
2. tube feet.
3. skin gills or dermal branchiae.
4. an inner calcareous endoskeleton that may have been the forerunner of the vertebrate skeleton, and
5. pincerlike pedicellariae which help keep the body free of debris, aid in capturing food and protect the skin gills.

You will study an adult starfish (see Fig. 13–1) to observe the basic plan of organization of an echinoderm. Place a starfish in a dissecting pan with its lower or oral surface facing upward. The mouth and tube feet are located on the oral surface. Five arms, each essentially alike, radiate outward from a central disk. Place the dissecting pan on the stage of a stereoscopic microscope and examine the tube feet. Turn the starfish over and examine the aboral surface. Identify the numerous calcareous spines. Located among the larger spines are very tiny pincerlike pedicellariae and soft skin gills. What is the function of the pedicellariae? _____ The pincerlike pedicellariae help to protect the very delicate skin gills. In the central disk region is the round sievelike madreporite, which is the entrance into the water vascular system. The very small anus also is located in the central disk region. Transfer the dissecting pan from the microscope stage back to the lab table.

Remove the aboral surface of one arm and of the central disk with a pair of scissors being careful to leave the madreporite. Located nearest to the aboral surface is a pair of digestive glands or hepatic ceca which secrete enzymes into the sac-like stomach. A pair of gonads (testes or ovaries) is located beneath the digestive glands. Sexes are separate in starfish and fertilization is external.

Next, locate the parts of the water vascular system. Water enters the sievelike madreporite and flows downward through the stone canal to the ring canal that encircles the mouth. A radial canal leads into each arm from the ring canal. Each tube foot is connected to a radial canal by a short lateral canal with a valve. Each tube foot consists of an upper expanded ampulla containing circular muscles and a lower sucker containing longitudinal muscles. Alternate contraction of the circular and longitudinal muscles creates and releases suction in the tube feet enabling the organism to pull itself along. The same principle is used in pulling apart the shells of bivalves on which the starfish feeds. Remove one tube foot and examine it with the stereoscopic microscope.

Examine the demonstration material pertaining to the echinoderms. Five classes are recognized. Adaptive radiation of body form and dietary specialization make them fairly diverse.

Class Asteroidea—starfish, sea stars; robust arms not well set off from a central disc, slowly motile, carnivorous on bivalves

Class Ophiuroidea—brittle stars, serpent stars; slender arms well set off from the central disc, rapidly motile, carnivorous on crustacea

Class Echinoidea—sea urchins, sand dollars; no free arms, long spines, skeleton fused into a shell, motile, primarily herbivorous

Class Holothuroidea—sea cucumbers; elongated body, no free arms, spines reduced so body covering feels leathery, motile, feed on dead organic matter in sand

Class Crinoidea—sea lilies, five free arms branched into pinnules that impart flower-like appearance, sessile, filter feeders

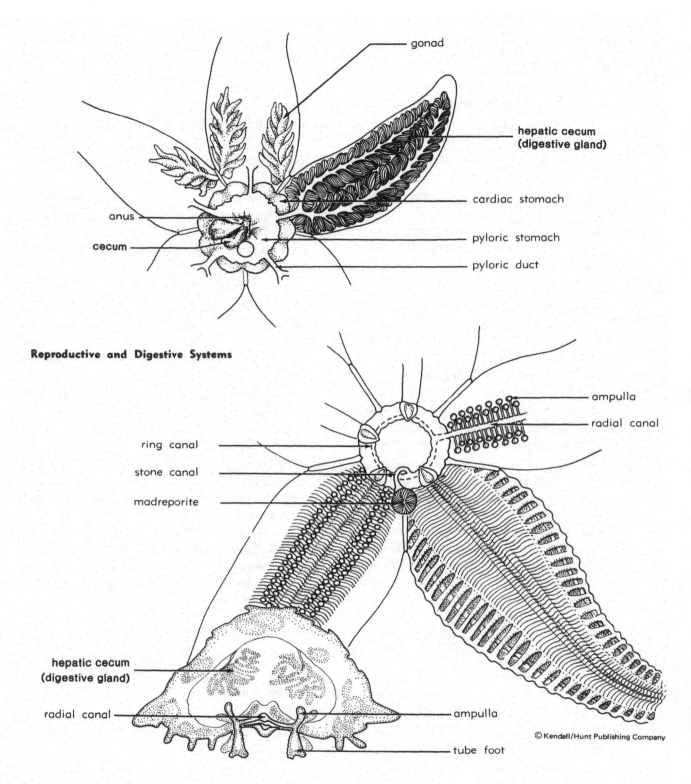

gonad

hepatic cecum
(digestive gland)

cardiac stomach

pyloric stomach

pyloric duct

anus

cecum

Reproductive and Digestive Systems

ampulla

radial canal

ring canal

stone canal

madreporite

hepatic cecum
(digestive gland)

radial canal

ampulla

tube foot

© Kendall/Hunt Publishing Company

Water Vascular System

Figure 13-1.

III. Phylum Chordata

All species of chordates exhibit the following characteristics at some stage in their life cycle.

1. An internal supporting rod or notochord. The notochord is replaced by a backbone in the subphylum vertebrata.

2. A dorsal, hollow nerve cord.

3. Pharyngeal gill slits.

Observe the three characteristics listed above on a slide of a whole mount of *Amphioxus*. The same three characteristics should be observed in a slide labeled *Amphioxus* pharynx c.s. Refer to Figure 13–2.

Amphioxus is a member of the subphylum Cephalochordata (lancelets) (16 species) and is an example of an invertebrate chordate. It is believed that *Amphioxus* is close in appearance to the primitive ancestor of this phylum. *Amphioxus* is common in shallow areas and one species is found in the Chesapeake Bay. In China they are so abundant that tons are harvested as food.

Another Chordate subphylum is Urochordata (1,300 species), consisting of marine organisms with tadpolelike larvae and saclike stationary adults covered by a tough outer covering. The third chordate subphylum is Vertebrata (41,700 species). Vertebrates are characterized by the possession of a vertebral column and the development of a protective cranium for the brain.

IV. Variations among the Vertebrates

The vertebrates are subdivided into seven living classes:

1. Agnatha [jawless fish]
2. Chondricthyes [cartilaginous fish]
3. Osteichthyes [bony fish]
4. Amphibia [amphibians]
5. Reptilia [reptiles]
6. Aves [birds]
7. Mammalia [mammals]

Examine the specimens from each class and describe those characteristics that have enabled each organism to adapt to its environment. Note the evolutionary trends in limb specialization, number of gill slits, and type of body covering.

A. CLASS AGNATHA [hagfish and lampreys]

Examine the lamprey specimen, sketch the overall shape of the body and describe:

1. the number and position of fins _____

2. the shape and structure of the mouth _____

3. the number and position of gill slits _____

4. the texture of the body covering _____

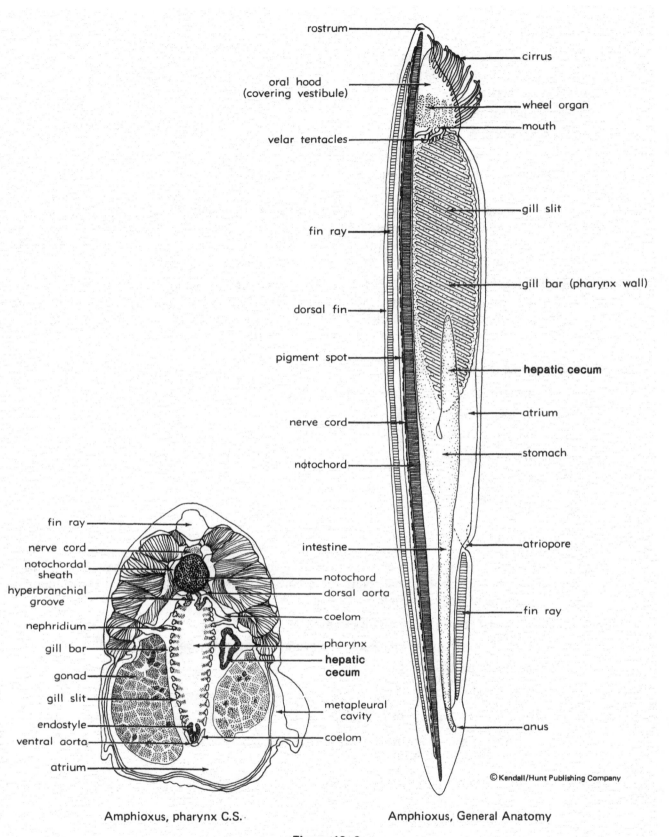

rostrum

cirrus

oral hood
(covering vestibule)

wheel organ

mouth

velar tentacles

gill slit

fin ray

gill bar (pharynx wall)

dorsal fin

pigment spot

hepatic cecum

atrium

nerve cord

stomach

notochord

fin ray

nerve cord

notochordal
sheath

notochord

dorsal aorta

hyperbranchial
groove

coelom

intestine

atriopore

nephridium

pharynx

gill bar

**hepatic
cecum**

gonad

gill slit

metapleural
cavity

fin ray

endostyle

coelom

ventral aorta

anus

atrium

© Kendall/Hunt Publishing Company

Amphioxus, pharynx C.S.

Amphioxus, General Anatomy

Figure 13-2.

Based upon your observations, infer the most probable type of swimming motion for the lamprey.

How is the mouth structure well adapted for an ectoparasitic lifestyle?

Examine Figure 13–3 and label the unpaired median fins and the seven gill slits. Also label the circular sucking disk, located around the mouth, and the horny teeth. The parasitic lamprey attaches to the surface of a host with the sucker and tears an opening through the skin with the horny teeth. Blood and soft tissue then are sucked by the lamprey.

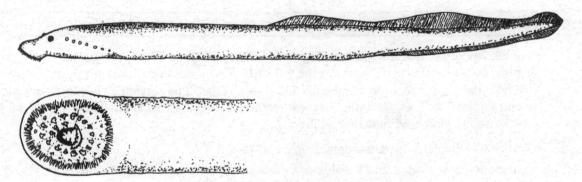

Figure 13–3. Lamprey.

In order to reproduce, lampreys must swim (in a wriggling fashion) up into shallow streams. Fertilized eggs develop into a larval form called *ammocoetes*. Examine a whole mount slide labeled *ammocoetes* under low magnification. What other chordate does it resemble

_____

Sketch the *ammocoetes* larva and label the three major chordate characteristics.

B. CLASS CHONDRICTHYES [sharks, rays, skates, chimerae]

Obtain a dogfish shark and carefully examine the body covering. Contrast it with the smooth skin of the lamprey.

Examine the slide, *Dogfish Placoid Scales, w.m.* or place a thin slice of skin under the microscope. The spine of each scale is covered with a tough enamel-like substance under which is dentine making up the bulk of the scale. In the center of each scale is a pulp cavity filled with blood vessels. The composition of the placoid scales is similar to what structures in your body?

_____ Next examine the mouth and describe how it differs from that of the lamprey.

Notice the several rows of teeth embedded in the flesh. From what structures have the teeth evolved?

_____

Count the number of pairs of gill slits. Now find a small opening just posterior to the eye. This is a remnant of the first gill slit and is called a spiracle. In bottom dwelling rays, most of the water passing over the gills enters through the spiracle. Observe the cartilaginous skeleton on demonstration and identify the gill arches.

Examine and identify the fins on the specimens. The paired lateral appendages allow for greater efficiency in locomotion. In what ways does this adaptation allow for a predatory lifestyle?

Another adaptation for a successful predatory lifestyle is the presence of an efficient digestive system. What advantages are obtained by the shark's enormous liver, large stomach and intestines, and a special spiral shaped partition located along the inner length of the intestines?

C. CLASS OSTEICHTHYES [bony fish]

Compare and contrast the perch specimen with that of the shark in reference to:

	BODY COVERING	GILLS	FINS
SHARK:			
PERCH:			

Note the smooth scales and gill covering called the operculum. The operculum not only protects the gills, but can help retain water in the gills if the fish is temporarily removed from the water. Also note the higher position of the gills and pectoral fins. The pelvic fins are more anterior and an anal fin is present. Observe the fin construction on the perch skeleton. The finer ray construction allows for increased maneuverability.

D. CLASS AMPHIBIA [salamanders, frogs, and toads]

Fossil evidence indicates that amphibians evolved from ancestral bony fish that possessed muscular lobed fins instead of ray fins. Only seven species of lobed-finned fish exist today compared to over 50,000 species of bony fish. What do these figures imply about the competition between ray-finned and lobed-finned fish?

_____

Observe the amphibian specimens and note the position of the two pairs of limbs. Do they extend down from the body or out to the side?

Due to lateral limb placement, many amphibians, such as salamanders, must wriggle their body from side to side as they move. In what way has the frog adapted to land movement?

_____

Contrast the skin of an amphibian with that of a fish. Is a thin moist skin an advantage or disadvantage on land? Under the skin are many blood vessels which allow for gas exchange through the thin skin. What does this imply about the efficiency of amphibian lungs?

The gas exchange mechanism is varied among the amphibians. Many amphibians utilized gills during the larval aquatic stage before metamorphosing into terrestrial adults. Some salamanders, such as the mud puppy, still utilize external gills during their entire life.

E. CLASS REPTILIA [turtles, crocodiles and alligators, lizards and snakes]

Reptiles are thought to have evolved from amphibian ancestors called Labyrinthodonts. The reptiles are the first vertebrates to be able to live entirely out of water. (Amphibians must return to the water to reproduce.) Observe the reptile specimens and describe the body covering, position of the limbs, and specialization of the digits. In what ways are each of these adaptive for land existence?

Other major adaptations for successful land existence include:

1. conservation of water through the ability to excrete nitrogenous wastes in the form of insoluble uric acid
2. mating on land by means of internal fertilization
3. development of a land egg with a leathery or limy shell to prevent drying out and a series of internal extra embryonic membranes for gas exchange, absorption of stored nutrients, storage of metabolic wastes, and bathing and cushioning the embryo

F. CLASS AVES [birds]

One of the limitations of many reptiles is the inability to maintain a constant body temperature. Recent investigations have indicated that some reptiles (now extinct) may have been homeothermic and could regulate their internal temperature. One group possessed featherlike structures which insulated the body. From this group evolved the birds. Observe the birds on demonstration. What two clues to their reptilian ancestry may be found covering the legs and toes? _____

_____

Examine the skeleton of a bird. Identify the following adaptations for flight:

1. The jaws have been modified into light horny beaks lacking teeth. This adaptation also allows for the elimination of heavy jaw muscles, thus keeping the center of gravity far back in the body.
2. The forelimbs have been modified into wings. The tip of each wing consists of several fused bones which provide added strength and rigidity.
3. The forelimbs are attached to the rest of the skeleton by means of a pectoral girdle. Locate the two clavicles which join together to form the "V" shaped "wish bone."
4. Locate the large median keel to which powerful flight muscles attach.
5. The shortened tail is well adapted for use as a rudder, stabilizer and brake.

If available, observe bird specimens on display. Note the variations of beaks and feet. For each specimen observed, sketch and/or describe the beak and feet.

Specimen No.	Beak	Feet

Identify the specimens that are adapted for:

—probing into small openings for insects: _____

—breaking open tough seeds: _____

—cutting and tearing flesh: _____

—holding slippery fish: _____

—swimming: _____

—perching on branches: _____

—grasping prey: _____

G. CLASS MAMMALIA [mammals]

Mammals evolved from a different group of reptiles than gave rise to the birds. Mammals are homeothermic. Gas exchange involves lungs assisted by a muscular diaphragm. Some mammals lay eggs but most bring forth living young. Mammals are family animals, spending a great deal of time caring for their young. Distinguishing features of the group include:

1. A covering of hair sometime during their life.

2. Mammary glands.

3. A well-developed nervous system.

4. Differentiation of teeth.

The one factor above all others that has enabled mammals to flourish and to adapt themselves to a wide variety of habitats is their highly developed nervous system.

Observe the demonstration displays of mammals.

Twelve naturally occurring orders of mammals are represented in the United States and coastal waters of the United States. Alphabetically arranged the orders are:

Artiodactyla —bighorn sheep, deer, elk, moose, pig and pronghorned antelope
Carnivora —bears, bobcat, badger, foxes, mountain lion, otter, raccoon, skunks, weasels and wolf
Cetacea —porpoises and whales
Chiroptera —bats
Edentata —armadillo
Insectivora —moles and shrews
Lagomorpha —hares, pikas and rabbits
Marsupialia —opossums
Pinnipedia —seals and sea lions
Primates —humans
Rodentia —beavers, mice, muskrat, porcupine, rats and squirrels
Sirenia —manatee

Both the fossil forms and modern forms exhibit adaptive radiation with an increase in number of species and specialization to fill numerous niches. Some of the evolutionary trends in the mammals include:

1. Change from egg laying habit to live bearing

2. Reduction of bones in the appendages

3. Reduction in total number of teeth

4. Change from non-specialized teeth (all similar) to specialized teeth not all similar in shape and function

The number and kinds of teeth (dentition) provide valuable information for mammalogists. There are four types of teeth in mammals and they are:

a. incisors—in front of mouth; have a beveled cutting edge used for cutting and gnawing food

b. canines—sharp pointed teeth for tearing, piercing and shredding food

c. premolars—may be flat for grinding or with pointed cusps for tearing

d. molars—in back of mouth; large and flat in herbivores for grinding food, more pointed in carnivores

Adult humans have four incisors, two canines, four premolars, and six molars in each jaw for a total of 32 teeth. The dental formula for each half jaw is written $\dfrac{2\text{-}1\text{-}2\text{-}3}{2\text{-}1\text{-}2\text{-}3}$ Humans are generalists or omnivores and no tooth type dominates the others. Herbivores tend to have large flattened molars and premolars, poorly developed canines and well developed chisel shaped incisors.

Carnivores have poorly developed incisors and sharply pointed canines, premolars and molars.

Examine the mammalian skulls on display. Note the number of teeth, the kinds of teeth, and determine the dental formula for each species. Judge whether the animal is a carnivore, herbivore or an omnivore. Record your findings in the following table.

Skull	Species	Order	Total Number of Teeth	Dental Formula	Probable diet	Notes
1						
2						
3						
4						
5						
6						

Which is the most primitive (unspecialized)? _____

Which is the most specialized _____

REVIEW QUESTIONS

1. In deuterostomes the first opening in the embryo becomes the _____ .

2. Identify five main features of echinoderms which distinguish them from other animal phyla.

(1)

(2)

(3)

(4)

(5)

3. Trace the flow of water through the water vascular system of a starfish and describe two of its functions.

4. List the three unique characteristics exhibited by all chordates at some stage in their life cycle.

   (1)

   (2)

   (3)

5. In *Amphioxus,* the supportive structure located just beneath the hollow nerve cord is the
   _____.

6. What evidence was presented in the laboratory exercise that supported the evolutionary relationship between the cephalochordates and the vertebrates?

7. Compare and contrast the following individuals in reference to gill slits, fins and body covering:

   (1) lamprey

   (2) shark

   (3) perch

8. Describe the adaptations and limitations for life on land possessed by amphibians and reptiles.

9. Describe at least four features of bird evolution that are adaptations for flight.

   (1)

   (2)

   (3)

   (4)

10. Match the items below with the appropriate vertebrate class.

_____ feathers

_____ hair

_____ no jaws

_____ seven gill slits

_____ specialized teeth

_____ an operculum

_____ bony pectoral, pelvic and anal fins

_____ hollow bones

_____ cartilaginous skeleton and placoid scales

A. Agnatha
B. Chondricthyes
C. Osteichthyes
D. Amphibia
E. Reptilia
F. Aves
G. Mammalia

ANIMAL TISSUES

Objectives

Upon completion of this exercise you should be able to:

1. Define the term "tissue" and list the four basic animal tissues.
2. Match each tissue with its structural characteristics.
3. List the function of each tissue studied.
4. Cite examples from humans where each tissue may be found.
5. Identify and label the component parts of each tissue type from a slide or diagram.
6. Differentiate among the various classifications of epithelial tissue based on shape and arrangement.
7. Describe three surface specializations of epithelial tissue, give the function of each and name a specific structure in the body where each is found.
8. Differentiate among the various types of connective tissue in terms of structure and function.
9. Explain the relationship between cells and nonliving intercellular material for each tissue group.
10. Differentiate among the three major types of muscle tissue.
11. Answer all questions in the exercise.

I. Introduction

At one stage or another, all organisms are essentially composed of just a single cell. Some organisms such as the ameba, and the green alga *Chlamydomonas,* remain at the one-cell stage through their entire life. In this case, the individual cell must be able to perform all of the basic properties of life. It must be able to:

1. Obtain and process nutrients.
2. Transport nutrients to all parts of the organism.
3. Exchange gases (O_2 and CO_2) with the environment.
4. Eliminate wastes.
5. Maintain proper salt and water balance in the environment.
6. Provide a protective boundary between the organism and the environment.
7. Sense changes in the environment.
8. React to environmental changes.
9. Reproduce.

All of these processes are performed by all single-cell organisms.

Multicellular organisms must also perform all of the processes mentioned above. They do this, not by having *each* cell accomplishing every task, but by having certain groups of cells specialize to perform a specific task with a high degree of efficiency. These groups of cells are called tissues. Therefore, a *tissue* is a group of cells all of which are similar in structure and which act together to perform a specialized function.

II. Prerequisite

You should have successfully completed the unit on the use of the microscope before you begin this unit.

All of the tissues you will study in this unit are grouped into four major types:

1. Epithelial.
2. Connective.
3. Muscle.
4. Nervous (discussed further in a later laboratory).

III. Epithelial Tissue

A. Observation of Epithelial Tissue.

1. Obtain the following three prepared slides:

    a. *Simple squamous epithelium w.m.*

    b. *Simple cuboidal epithelium* (thyroid or kidney)

    c. *Goblet cell of columnar epithelium.*

2. Observe simple squamous epithelium first under low power and then under high power. Draw a small section of the field (about five cells). Note that the tissue consists of a single layer of flattened cells.

3. Next view the slide labeled simple cuboidal epithelium under low power. In the thyroid the field of vision should consist of many pink stained areas. Each of these pink areas is surrounded by epithelial cells. Draw one of these pink areas. If you use the kidney look for long longitudinal structures, the ducts, and tubules, which are comprised of epithelial cells. Draw several cells from one of the ducts or tubules. Simple cuboidal epithelium consists of a single layer of cells with each cell about as tall as it is wide.

4. Examine the slide of columnar epithelium under low power. Adjacent to the lumen of the intestine find a single layer of cells taller than they are wide. Switch to high power and draw a section showing at least five epithelial cells. (Note: the large specially stained oval structures represent mucus contained within goblet cells.)

B. Function and Location of Epithelial Tissue.

1. Which one of the following statements best characterizes all of the epithelial tissues you have seen?

_____ a. The cells of epithelial tissue are dispersed among a large amount of intercellular material.

_____ b. The cells of epithelial tissue are located very close together with very little intercellular material.

In each diagram you have made, the epithelial cells have been very close together. The only material between them is a very small amount of cementing substance. In addition, one edge of epithelial tissue borders on an open area. This arrangement allows epithelial tissue to serve very well as a *boundary* between the organism and the environment or to serve as a lining within the organism.

2. Epithelial tissue also functions in:

a. absorption of substances from the outside medium

b. elimination of substances to the outside

3. Therefore epithelial tissue may be found lining:

   a. the outer surface of the body (skin, scalp)

   b. the digestive tract (mouth, esophagus, stomach, etc.)

   c. the kidney tubules

   d. the lungs

   e. the genital and excretory tracts

4. Check those structures which are lined by epithelial tissue:

   _____ a. trachea (windpipe)

   _____ b. small intestine

   _____ c. nasal cavity

(All three should have been checked)

C. Classification of Epithelial Tissue.

   1. Epithelial tissue may be subdivided on the basis of the shapes of the cells. The three basic shapes of cells are:

      a. *squamous*—thin, flattened cells with no specific outline

      b. *cuboidal*—the height is about the same as the width

      c. *columnar*—the height is much greater than the width

   2. Next to each diagram in figure 14–1, indicate the type of epithelial tissue based on the shape of the cells.

   3. Epithelial tissue may also be classified based on the arrangement of cells (i.e., the number of cell layers present). The two basic arrangements of cells are:

      a. *simple*—only one layer of epithelial cells present

      b. *stratified*—more than one layer present with only the lowest layer touching the underlying tissue

   4. Refer back to Figure 14–1. Amend each label to include the classification based on shape *and* arrangement of cells. Note that in classifying a stratified section on the basis of cell shape, the name is determined by the type of cell which borders the free (distal) surface. For example, tissue section "D" should be classified as "stratified *squamous* epithelial tissue." Examine stratified squamous in a prepared slide of the skin.

D. Specializations of Epithelial Tissue.

   The *distal surface* (farthest away from the underlying tissue) of those cells which are located at the free end of epithelial tissue may be further specialized.

   1. *Microvilli*—These are minute fingerlike projections on the distal end of epithelial cells lining the small *intestine* and on the proximal convoluted tubules in the *kidney*. Their function is to *increase the surface area* for absorption.

161

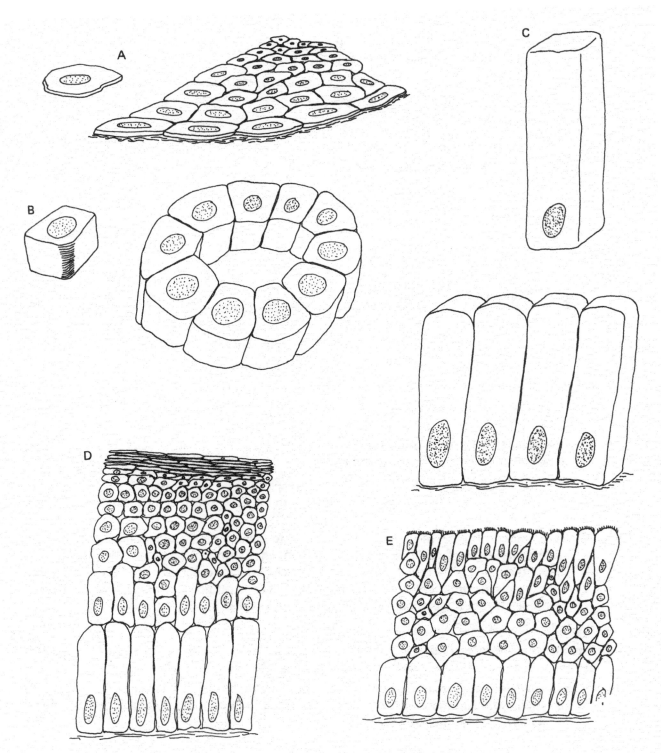

Figure 14-1.

a. Examine the slide of *small intestine—mammal c.s.* under low power. Locate the *villi* (folds) which project into the lumen of the intestine (Fig. 14–2).

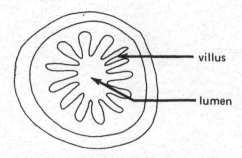

villus

lumen

Figure 14–2.

b. Now place a single villus under high power.

    (1) Which of the four basic animal tissues lines the intestine? _____

    (2) Based on arrangement, is this epithelial tissue simple or stratified? _____

    (3) Is this simple tissue squamous, cuboidal or columnar? _____

c. Look very closely at the distal surface of these columnar cells. By careful focusing, you should see that the extreme edge of the cell appears as a thin dark line. This is called a *striated border* and appears dark because of all the microvilli present. (Fig. 14–3.)

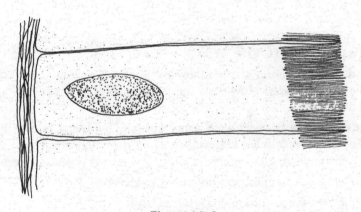

Figure 14–3.

2. Cilia.

    a. Place the slide *ciliated columnar epithelium* or the slide *mammalian trachea* under low power. Locate the free edge (that which lines the lumen of the trachea). Center this area and switch to high power. At the distal end of the columnar cells are clumps of *cilia* and *mucus*. Make a *sketch* in the margin of the distal surface showing cilia and mucus.

    b. Dust particles which are inhaled stick to the mucus and as the cilia move in rhythmical fashion, the dust-laden mucus is directed toward the oral cavity where it is swallowed.

    c. Ciliated epithelium is also found in the testes, oviduct, uterus and nasal sinuses.

3. *Cuticle*—In mammals the epithelial cells of the teeth, eye lens and internal ear secrete a material which covers the entire free surface of the tissue. In the tooth this cuticle becomes impregnated with various salts and is known as enamel.

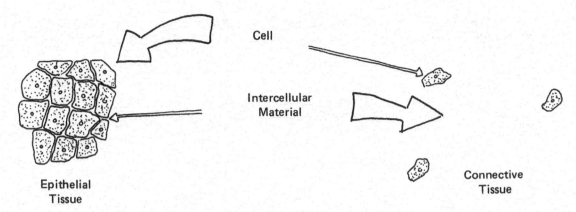

Figure 14-4.

Epithelial Tissue

Cell

Intercellular Material

Connective Tissue

IV. Connective Tissue

Both epithelial tissue and connective tissue are composed of *living cells* and *nonliving intercellular* material. In epithelial tissue, the amount of intercellular material produced was just enough to cement the cells together. This is not the case in connective tissue. The volume of intercellular material produced *greatly* exceeds the volume of cells (fig. 14-4).

Connective tissue may be subdivided into:

    a. blood (blood cells and nonliving plasma) (will be studied during future labs)

    b. connective tissue proper (CTP)

    c. cartilage

    d. bone

A. Connective Tissue Proper (CTP).

1. Examine a slide of *areolar tissue* under low power. The small dark staining "dots" represent *nuclei* of the CTP *cells*. How does the arrangement of these cells differ from that of epithelial cells?

2. The large clear area between cells represents the nonliving intercellular material. Running throughout the intercellular material in every direction are *fibers*. Fibers are produced by the CTP cells and are characteristic of CTP. Several types of fibers are found in CTP:

    a. *collagenous fibers* (flexible, but resist stretching)

    b. *elastic fibers* (very thin and stretch easily)

    c. *reticular fibers* (exhibit extreme branching and are found primarily where connective tissue meets another type of tissue. For example, the reticular fibers form a *basement membrane* at the junction of epithelial and connective tissue).

3. Draw a section of areolar tissue under high power and label: *cells, intercellular material,* and *fibers*.

4. CTP contains a variety of cell types. Many of the cells are relatively undifferentiated but can later give rise to more specialized types. Certain cells function in storing fat. Fat droplets enter the cell and push the cytoplasm out to the edge of the cell where it appears only as a thin membrane. Many of these fat cells tend to accumulate in large numbers and gradually crowd out the remaining cells. At this point, the tissue is referred to as *adipose tissue*. Obtain the slide: "*adipose tissue*". Examine it under high power and draw a cell showing *nucleus, cytoplasm* and the large empty space where a *fat droplet* used to be.

5. The *intercellular material* (ground substance) is a complex mixture of proteins, carbohydrates, lipids, and water.

6. Refer back to the slide of areolar tissue. Check which *one* of the following *best* applies.

    _____ a. the fibers are loosely arranged.

    _____ b. the fibers are very densely arranged in thick bundles.

    Due to the arrangement of fibers, this type of tissue is called *loose connective tissue*.

7. Loose connective tissue:

    a. provides support for other tissues.

    b. allows proper movement of connective structures.

    c. forms thin sheets which surround tissues.

    d. fills in spaces between organs.

    e. contains cells which can engulf foreign particles.

8. Just beneath the epithelial tissue of the skin, intestine, and kidney, the CTP is very dense and is called *dense connective tissue*.

9. Some CTP possesses fibers which are arranged in a definite regular order.

    a. Observe the slide: *Muscle-tendon* and view it with the naked eye. Most of the tissue on the slide is muscle tissue. At one end, however, is *dense* connective tissue (the *tendon*) which serves to connect the muscle to a bone.

    b. Now observe the tendon portion under low and high power. The fibers offer great resistance to a pulling force. Therefore, most of the fibers would probably be _____ (collagenous, elastic, reticular).

10. Dense connective tissue may also be found in:

    a. ligaments (connect bone to bone)

    b. the sclera (white) of the eye

    c. the cornea (transparent)

    d. the tough covering of organs

B. Cartilage.

1. Obtain a slide of *hyaline cartilage*. Being connective tissue, cartilage possesses two basic characteristics:

   a. living _____

   b. nonliving _____

   The living cells secrete a nonliving matrix which gives cartilage its flexibility. The cells are located within small spaces called *lacunae*. Draw a section of hyaline cartilage and label *cell, lacuna* and *matrix*.

2. Some cartilage may contain obvious fibers such as *elastic* or *fibro*cartilage.

3. The embryonic skeleton is composed of cartilage. Gradually it is replaced by bone except in certain areas. These areas include:

   a. pinna (flap) of ear

   b. tip of nose

   c. end of ribs and sternum

   d. ends of bones at movable joints

C. Bone is similar to cartilage except that the matrix contains calcium and phosphate salts which make it very hard.

1. Observe the slide: *bone ground human*. The very dark areas are the spaces (Haversian canals) where blood vessels, and nerves used to be located. The very small dark areas located around this central canal are the lacunae where the bone cells used to be located. In between are concentric layers of intercellular matrix (lamella). The very thin lines (canaliculi) radiating from the lacunae to the central canal represent passage ways for nutrients and wastes.

2. Label the Haversian system (D) in Figure 14-5. Include the *Haversian canal, lacunae, concentric lamellae* and *canaliculi*.

V. Muscle Tissue

1. The major function of muscle cells is to *contract*. To accomplish this function, the muscle cell has become highly specialized. The cells are greatly elongated and are referred to as *fibers*. Many fibers are bundled together and wrapped in connective tissue.

   Which type of connective tissue performs this function?_____

_____

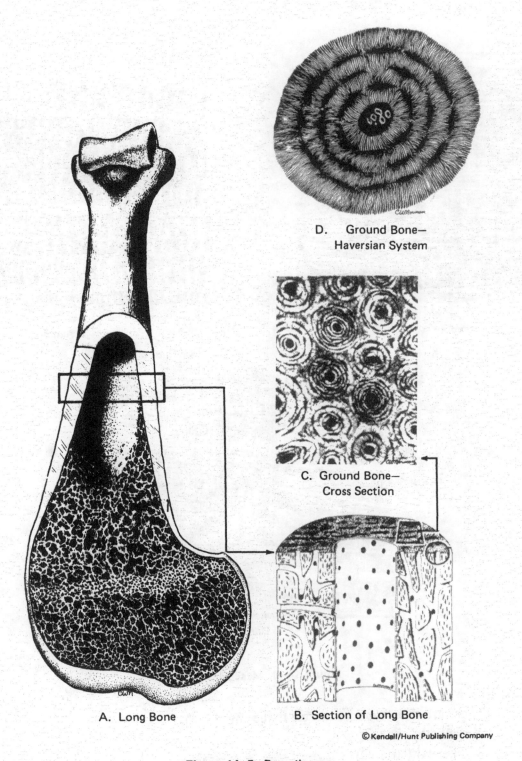

D. Ground Bone—
Haversian System

C. Ground Bone—
Cross Section

A. Long Bone

B. Section of Long Bone

Figure 14-5. Bone tissues.

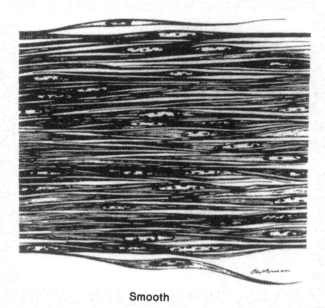

Smooth

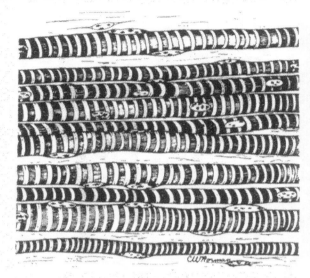

Skeletal

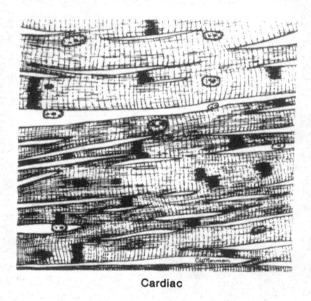

Cardiac

Figure 14–6. Muscular tissues.

2. There are three types of muscle tissue (fig. 14–6).
   a. smooth
   b. skeletal (striated)
   c. cardiac

3. Obtain the slide: *muscle types*. View the slide with your naked eye.

   Place the section (intestine) containing smooth muscle under low power. Lining the lumen of the intestine is _____ tissue. Moving from the lumen out past the simple columnar epithelial tissue, you next view several layers of dense _____ tissue. Past the connective tissue you arrive at the deep pink longitudinal smooth muscle tissue.

4. Switch to high power. Each elongated muscle fiber contains one nucleus. The ends of each fiber are extremely tapered. Smooth muscle is *not* under voluntary control. Therefore, in which *one* of the following would smooth muscle *not* be responsible for movement?

_____ a. wall of digestive tract

_____ b. wall of an artery

_____ c. bronchus

_____ d. biceps brachii

5. Switch back to low power. Move the slide so that the *field* passes to the *left,* so that the center section is in focus. Move the *field* down until you see longitudinal fibers running across the field. These are striated fibers of skeletal muscles. The striated fibers are very long and contain many peripheral nuclei. Note that the individual fibers remain separate from one another.

6. Switch to high power. Each fiber consists of many longitudinal *fibrils*. These fibrils give the muscle a striated appearance. In addition to the longitudinal striations, the fiber possesses cross (up and down) striations. Skeletal muscles are attached to bone by a regular connective tissue called _____ . *Skeletal* muscles are also known as *striated* muscles and are under voluntary control.

7. Now switch back to low power and view the last section on the slide. This type of muscle tissue is called _____ . Which two of the four statements below about cardiac muscle are correct?

_____ a. The fibers are long and appear to branch.

_____ b. The fibers are long and do not branch.

_____ c. The nuclei are only located on the peripheral edge of each fiber.

(Switch to high power)

_____ d. There are longitudinal and cross striations.

8. Where two fibers meet end to end in cardiac muscle a small dark line (intercalated disc) can be seen. The impulse for contraction can easily be transferred from one fiber to the next because of this arrangement. The contraction of cardiac muscle is automatic and is not under voluntary control.

VI. Nervous Tissue

The *neuron* is the nerve cell type that is actively involved in the transmission of nerve impulses, the major function of nervous tissue. Examine a slide labeled Motor Neuron or Spinal Cord, ox under low power. Locate a large nucleated cell with numerous extensions or processes off the main cell body. Some of the processes may be very long. The numerous nuclei you see around the neuron are those of neuroglial cells, another nervous tissue cell type. Sketch a neuron below.

REVIEW QUESTIONS

Epithelial Tissue

1. Name one characteristic *all* epithelial tissues exhibit. _____

2. Name three functions of epithelial tissue.   a. _____
                                                      b. _____
                                                      c. _____

3. Epithelial tissue may be classed based on

    a. _____ and

    b. _____

4. Name two types of epithelial tissues based on arrangement of cells.

    a. _____

    b. _____

5. Name three types of epithelial tissues based on shape of cells.

    a. _____

    b. _____

    c. _____

6. Name two areas of the body where epithelial tissue may be found.

    a. _____

    b. _____

7. Name three specializations of the distal surface of some epithelial cells; give their function and an example of where the tissue is found in the body.

    a. _____ _____ _____

    b. _____ _____ _____

    c. _____ _____ _____

1. Name the four *major* types of connective tissue:

   a. _____

   b. _____

   c. _____

   d. _____

2. Beside cells and intercellular material, what other structure do some connective tissues possess?

   _____

3. Name one place dense connective tissue is found. _____

4. What is a lacuna? _____

5. How does the matrix of bone differ from that of cartilage? _____

   _____

Muscle Tissue

1. Name the three types of muscle tissue.

   a. _____ b. _____ c. _____

2. Skeletal muscle is also known as _____

3. Which type of muscle is under voluntary control? _____

4. Which type of muscle is found in the wall of the esophagus? _____

5. What is the function of a tendon? _____

6. What type of tissue is a tendon? Be as specific as possible. _____

7. Which type of muscle cell branches? _____

8. Which type of muscle cell lacks definite cross striations? _____

9. Answer questions on demonstration (if present). _____

Exercise 15

VERTEBRATE ANATOMY: DIGESTIVE AND RESPIRATORY SYSTEMS

Objectives

Upon completion of this exercise you should be able to:

1. Define the terms listed in the section on definitions and identify them on a vertebrate.
2. Identify the following external parts from a fetal pig or fetal pig diagram: umbilical cord, urogenital papilla, mouth, oral cavity, nostrils, external ears, auditory canal, eye, head, neck, trunk, tail, anus, nipples and hoofs.
3. Distinguish male and female pigs by using external features.
4. Identify the following parts of the respiratory system from a fetal pig or fetal pig diagram: pharynx, epiglottis, glottis, larynx, trachea, bronchi, lungs and diaphragm.
5. Locate the parts of the respiratory system listed in objective 4 on models of a human.
6. Identify the following parts of the digestive system from a fetal pig or a fetal pig diagram: teeth, tongue, hard palate, soft palate, esophagus, stomach, liver, gall bladder, small intestine, pancreas, villi, large intestine, cecum, rectum and anus.
7. Locate the parts of the digestive system listed in objective 6 on models of a human.
8. Give functions for the parts listed in objectives 4 and 6.
9. Answer all review questions in the exercise.

I. Introduction

In this and other exercises, the anatomy of the fetal pig will be studied. The exercises will also examine selected aspects of the anatomy and physiology of humans.

Of all the mammals, except the other primates, the pig is probably as close as any to humans in general body structure. The animals you will use were removed when their mothers were slaughtered in the stockyards for food. Because hogs are sold by the pound, farmers find it profitable to sell the females when pregnant. U.S. Food and Drug Regulations classify the unborn young as part of the inedible internal organs. Consequently, educational use of the fetus is a method of optimizing resources.

Familiarize yourself with the list of definitions provided below before continuing with the exercise.

DEFINITIONS

Anterior(ly)	—Head end or toward the head end.
Posterior(ly)	—Tail end or toward the tail end.
Ventral(ly)	—Belly side or toward the belly.
Dorsal(ly)	—Back side or toward the back side.
Lateral(ly)	—Side or toward the side.
Medial(ly)	—Middle or toward the middle.

Distal(ly)	—Free end of a limb or projection or toward the free end of a limb or projection.
Proximal(ly)	—End attached to the body or toward the end attached to the body.
Horizontal plane	—Any lengthwise plane that cuts across the body at a right angle to the median plane.
Median plane	—The one plane which divides a bilaterally symmetrical animal into two mirror images or right and left halves. It is also called a median sagittal plane.
Sagittal plane	—Any plane parallel to the median plane.
Transverse section	—Any thin section which cuts across the body at a right angle to the long axis.
Longitudinal section	—Any thin section (horizontal, saggital) cut parallel to the long axis of the body.

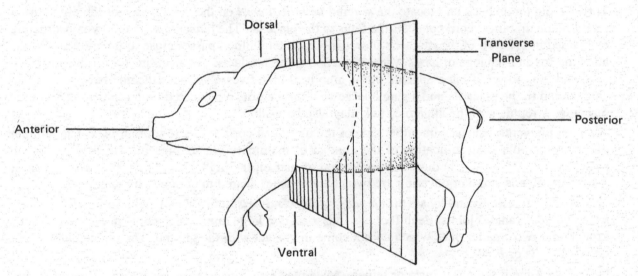

Figure 15–1A.

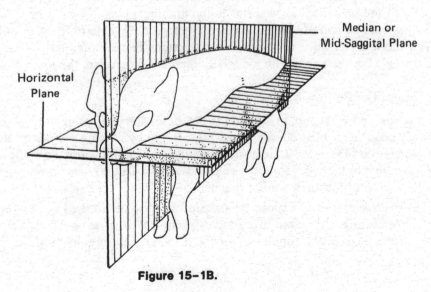

Figure 15–1B.

173

II. External Anatomy of the Fetal Pig

Students will study the pig in groups of two to four. Each group should obtain a male or female pig as directed by their instructor. Upon obtaining a pig, remove it from the plastic bag, wash it in running water and place it in a dissecting pan for observation. Note the slash on one side of the neck. This marks the location where red and blue liquid latex was injected into a major artery and vein. The latex became solid rubber within the pig, both strengthening the blood vessels and aiding in their identification.

Beginning at the anterior end, locate the *mouth*, which leads into the *oral* or mouth *cavity*, the *nose*, the *nostrils* (external nares), which lead into the *nasal cavity*, the *external ears*, and the external ear canal or *auditory canal*, which leads inward from the ears. Herbivores, such as the wild pig, are subject to prey by large cats and other predators. Consequently, they depend upon smell and hearing to detect the presence of their enemies, and the ears can be moved to enable better detection of the source of the sound.

Next, locate the *nictitating membrane* in the corner of the *eye*, the *eyeball*, *eyelashes* and *eyelids*. Now, lay the pig on its side and identify the major body divisions including the *head, neck, trunk* (which consists internally of an anterior thorax and a posterior abdomen separated by a muscular diaphragm), and *tail*. The *anus*, the posterior opening of the digestive tract, is located ventral to the tail.

Turn the pig on its back and locate the severed *umbilical cord* on the ventral side of the pig. During fetal life the umbilical cord connects the fetus to the placenta. The *placenta* is an embryonic structure embedded in the wall of the mother's uterus in which fetal blood passes close to the mother's blood allowing for an exchange of gases, water, food and waste products. Cut a centimeter off the umbilical cord and look to see if latex appears in the umbilical blood vessels. These include two arteries that carry blood to the placenta and a single large vein that returns oxygenated blood to the pig from the placenta. A remnant of the allantoic stalk may also be visible.

Also, on the ventral surface on either side of the umbilical cord, find a row of *nipples*. The number of nipples visible is an indication of the number of mammary glands present. Are they present in both sexes? _____ Is there any indication of hair on the pig? _____ The possession of mammary glands and hair covering are two characteristics shared in common by mammals.

Examine the forelimbs and locate the *shoulder, elbow* and *wrist* joints. On the hindlimbs find the *hip, knee* and *hock* (ankle) joints. Each foot has four toes, two large ones which come in contact with the ground and two smaller ones which do not come in contact with the ground. The pig actually walks on its toenails or *hoofs*.

Next, determine if the pig is male or female. Male pigs have a *penis* that lies entirely within the skin along the ventral posterior belly wall except during mating. The opening of the male urogenital tract at the end of the penis is just posterior to the umbilical cord base. Viewing the pig from behind, the *scrotum* appears as paired bulges. You are required to be able to identify both a male and a female pig even though you have only one for dissection. Female pigs have a single urogenital opening (located ventral to the anus) with a dorsally directed projection, the *urogenital papilla*, below it.

III. Dissection of the Pig

For dissection you will need, in addition to the dissection pan, string, a blunt probe, scissors, a scalpel or a single edge razor blade, forceps and pins. Begin by placing the pig on its back in the dissecting pan. Referring to Figure 15–2, tie a string around one forelimb and place the string around and under the pan and tie it to the other forelimb so that the pig is securely anchored and its forelimbs are spread apart. Tie the hindlimbs in a similar manner.

The goal of dissection is to expose the parts of the systems studied so that they may be examined. A minimum of cutting is required in a good dissection. The most useful tool is the dull probe. Use this blunt instrument to probe tubular organs and to separate organs along their natural planes. Most

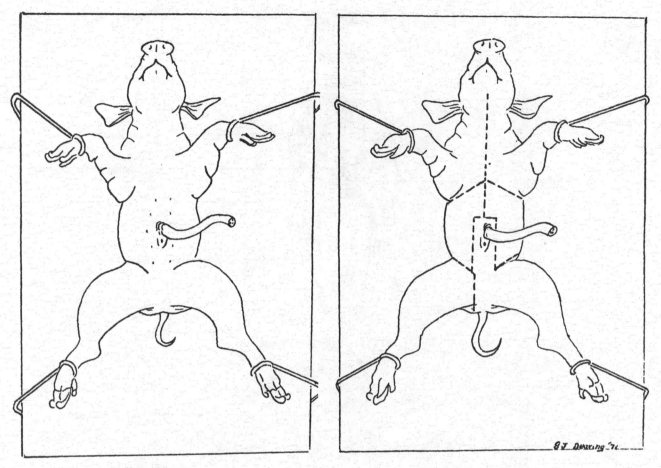

Figure 15-2.   Figure 15-3.

cutting should be done with scissors because they permit you to determine how deeply you are cutting. In some instances, a scalpel or a razor blade may be used to make an incision from the outside. As you proceed with the dissection, refer to the accompanying fetal pig diagrams. After each dissection slip the string from beneath the pan, return the pig to its plastic bag and place a tag on the bag identifying the members of your group.

IV. The Respiratory System (Figs. 15–4 and 15–5.)

This system facilitates the exchange of gases between the internal and external environments of the organism. It consists of the lungs and all passageways leading to and from the lungs. Air enters and leaves through the nostrils or external nares. These lead, via the nasal cavity and internal nares, into the *pharynx* or throat, the posterior continuation of the oral cavity. Use a blunt probe and try to follow this channel into the pharynx. Now expose the pharynx by inserting scissors into the corners of the mouth and cutting the jawbones. Find a cartilaginous projection, the *epiglottis*, at the base of the tongue. This flap aids in preventing food from entering the air passageways leading to the lungs and covers an opening, the *glottis*, which leads into the *larynx*, and on into the windpipe or *trachea*.

Using scissors, cut along the dotted lines as shown in Figure 15–3. To enter the thoracic and abdominal regions of the pig you will pass through an outer layer of skin, a muscular wall, and an inner shiny membrane. The membrane lining the abdominal wall is called *peritoneum*. The thorax wall is lined by *pleura*. You should also cut through the ribs. A vein, the *umbilical vein*, passes from the base of the umbilical cord to the liver. After identifying it, cut it, leaving both ends so they can be identified later.

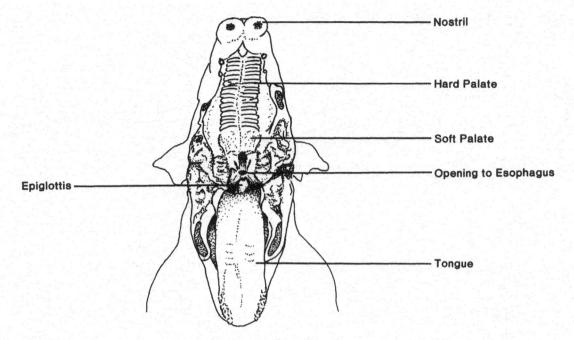

Nostril

Hard Palate

Soft Palate

Opening to Esophagus

Epiglottis

Tongue

A. Oral Cavity and Pharynx.

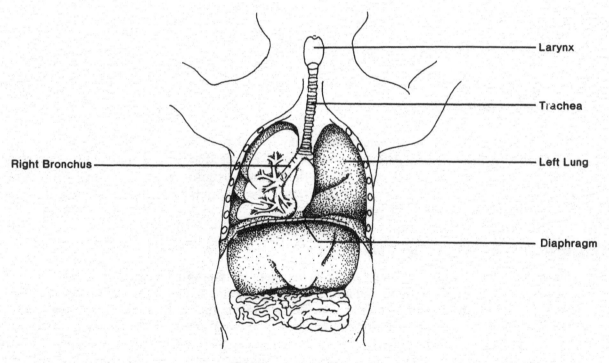

Larynx

Trachea

Right Bronchus

Left Lung

Diaphragm

B. Airways and Lungs.

Figure 15–4. Fetal pig, respiratory system.

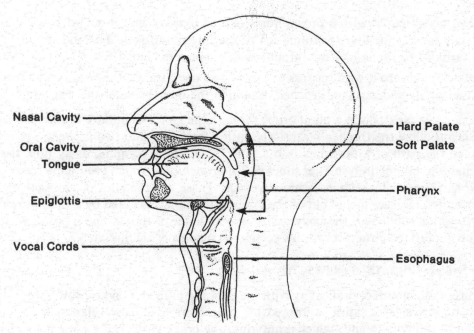

Nasal Cavity

Oral Cavity

Tongue

Epiglottis

Vocal Cords

Hard Palate

Soft Palate

Pharynx

Esophagus

A. Upper Respiratory System

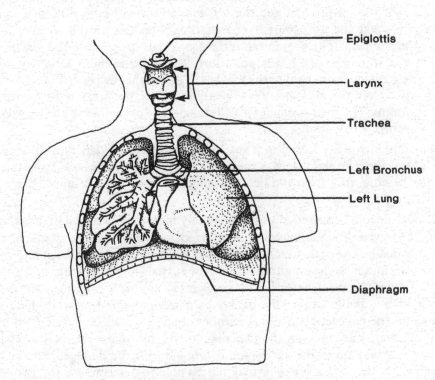

Epiglottis

Larynx

Trachea

Left Bronchus

Left Lung

Diaphragm

B. Lower Respiratory System.

Figure 15-5. Human, respiratory system.

If the body cavity is obscured with a brown fluid (clotted blood), rinse it out with tap water. Now, pin back the skin flaps to facilitate examination of the internal organs. The abdominal organs are held in place partially by the *mesentery* through which numerous blood vessels ramify.

Find the *larynx* and cut into it with scissors to expose two membranes, the *vocal cords*. Posterior to the larynx is the cartilaginous ringed *trachea*. Examine the dorsal side of the trachea. Is cartilage present? _____ Next, place a blunt probe through the glottis and move it until you locate the tip in the trachea. Follow the trachea posteriorly until it branches into two *bronchi*, one leading to each *lung*. Trace the path of a bronchus into a lung by carefully scraping away lung tissue. The bronchus continues to branch into smaller tubes called *bronchioles*, which eventually lead to tiny air sacs or *alveoli* surrounded by numerous blood vessels. These are the sites of gas exchange in the lungs. The movement of gases in and out of the lungs in mammals is aided by a thin muscular wall, the *diaphragm*, which subdivides the body cavity into an anterior thorax and a posterior abdomen. Now, locate the underlined structures in this section on models of the human.

V. The Digestive System (Figs. 15–6 and 15–7)

Digestion begins in the mouth or oral cavity with the chewing of food. Food is broken up by the teeth and moistened by saliva. The saliva is formed by three pairs of salivary glands, the parotid, the submaxillary and sublingual glands which empty into the oral cavity. The *tongue* is located on the floor of the mouth cavity. Roll your tongue along the roof of your mouth. The *hard palate* forms the hard anterior part of the roof and the *soft palate*, the soft posterior part. Locate these in the pig. A crossover of food and air occurs in the pharynx. Food leaves the pharynx via the *esophagus* on its way to the stomach. The opening into the esophagus is dorsal to the glottis. Insert a probe into the opening and trace the esophagus posteriorly through the thoracic cavity. Locate the esophagus in the thoracic cavity just dorsal to the trachea. The esophagus penetrates the diaphragm and continues posteriorly to join the *stomach*. Now locate the *liver*, the large red-brown lobed organ posterior to the diaphragm. The liver produces a secretion, bile, containing bile salts which emulsifies fats in the small intestine.

The bile is temporarily stored in a small sac, the *gall bladder*. Find the gall bladder by lifting the right side of the liver. Now locate the *cystic duct* which carries bile from the gall bladder. It merges with the *hepatic duct* from the liver forming the *common bile duct* before connecting to the duodenum.

On the left side and covered almost completely by the lobes of the liver is the stomach. The anterior end of the stomach near the heart is referred to as the cardiac stomach, the posterior end is the pyloric stomach. The *pyloric sphincter* is located between the stomach and *small intestine*. Cut open the stomach and note the longitudinal folds or *rugae* which disappear when the stomach is distended. The portion of the small intestine into which the stomach empties is the duodenum. Now locate a long, whitish granular organ, the *pancreas*, posterior to and partially hidden by the stomach. The secretion of the pancreas contains a number of enzymes which act on food in the duodenum. Try to locate the duct leading from the pancreas to the duodenum. Does it merge with the duct leading from the gall bladder to the duodenum before entering the duodenum? The pancreatic duct may be difficult to locate. Therefore, you may not be able to determine if the two ducts merge. To the left of the stomach locate the *spleen*, a red-brown flattened organ which functions in destroying old red blood cells in an adult. The small intestine is highly coiled providing a good deal of surface area at which the digestion and absorption of food can take place. Microscopic fingerlike projections called

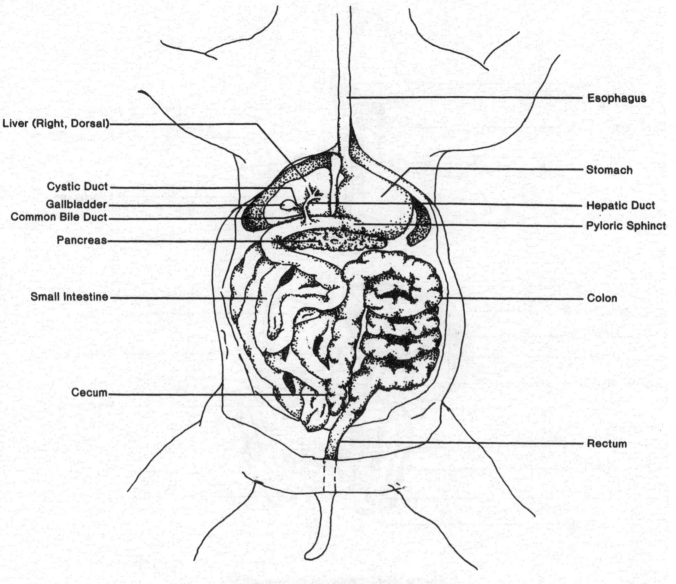

Figure 15-6. Fetal pig, digestive system.

villi line the small intestine, serving to increase surface area and to facilitate digestion. Snip out a short section of the small intestine, cut it open with scissors, and find the villi with a dissecting microscope. The small intestine continues posteriorly, eventually merging with the *large intestine,* or *colon.* Locate a blind pouch, the *cecum,* posterior to the junction of the small and large intestines. No appendix is present in the pig. (Note: In humans, an appendix is attached to the lower end of the cecum.) The large intestine loops around until it joins a straight section, the *rectum,* which opens to the outside by way of the *anus.* Now locate as many of the structures in this section as possible on models of the human.

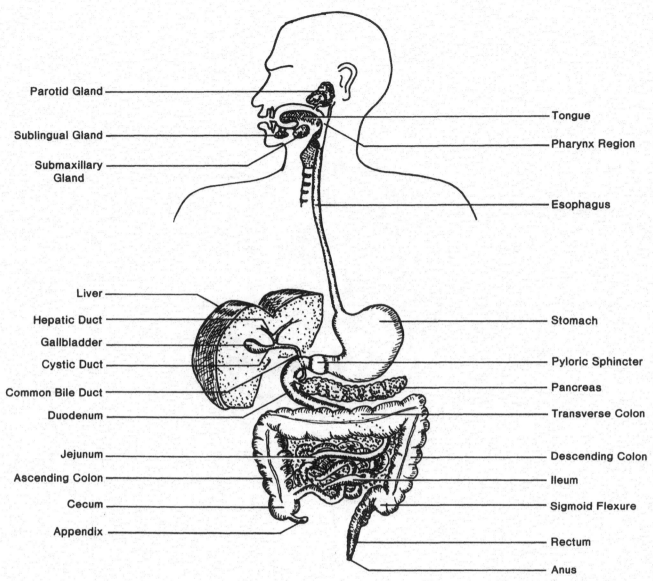

Parotid Gland

Sublingual Gland

Submaxillary
Gland

Tongue

Pharynx Region

Esophagus

Liver

Hepatic Duct

Gallbladder

Cystic Duct

Common Bile Duct

Duodenum

Jejunum

Ascending Colon

Cecum

Appendix

Stomach

Pyloric Sphincter

Pancreas

Transverse Colon

Descending Colon

Ileum

Sigmoid Flexure

Rectum

Anus

Figure 15-7. Human, digestive system.

REVIEW QUESTIONS

1. The part of the respiratory system located above the palate is the _____ .

2. The opening which leads into the larynx is the _____ .

3. The flaplike projection which covers the glottis is the _____ .

4. The trachea branches into two _____ each of which enters a lung.

5. The sites of gas exchange in the lungs are tiny sacs called _____ .

6. A thin muscular wall, the _____ , subdivides the body cavity into thorax and abdomen.

7. The tube leading from the pharynx to the stomach is the _____ .

8. The _____ secretes bile.

9. Bile is stored in the _____ .

10. _____ are fingerlike projections inside the small intestine.

Identify the following terms:

11. ventral— _____

12. anterior— _____

13. lateral— _____

14. proximal— _____

15. transverse section— _____

CHEMISTRY OF DIGESTION AND
PHYSIOLOGY OF GAS EXCHANGE

Objectives

Upon completion of this exercise you should be able to:

1. Define hydrolysis and dehydration.
2. Define digestion.
3. List the end products of digestion of carbohydrates, proteins, and lipids.
4. List the digestion events which occur in the mouth, stomach, and small intestine.
5. Describe or diagram the five principal methods of gas exchange.
6. Define tidal air, inspiratory reserve, expiratory reserve, vital capacity, total lung capacity and residual volume.
7. Determine your vital capacity, tidal volume, and expiratory reserve using the spirometer. Then compute your inspiratory reserve volume.
8. Compare and contrast aerobic respiration with gas exchange.
9. Determine the rate of CO_2 production in humans using a titration procedure.
10. Answer all review questions.

I. Digestion

All metabolism that occurs in an organism can be divided into two basic types of chemical reactions, one which builds molecules and the other which breaks down molecules. The cell is the basic unit of structure and function so that metabolism includes all of the chemical processes that occur in the cell. In order for cells to grow and reproduce, they need raw materials and energy. In multicellular organisms, such as the human being, the cells get the raw materials and energy from food. Most of this food is too complex to use directly and must undergo digestion (chemical breakdown) to simpler substances which then can be absorbed and used by cells. The digested food is used by each cell to make carbohydrates, proteins, lipids, and nucleic acids. During cellular respiration, energy, in the form of ATP, is produced by each cell from some of the digested food. Digestion for humans can be outlined as follows:

A. *Mouth.*

Mechanical breakup of food by means of teeth.
Food is moistened and lubricated by saliva.
Starch digestion begins in the presence of amylase (starch digesting enzyme) in saliva.

B. *Stomach.*

Further mechanical breakup of food due to muscular action of the stomach. Hydrochloric acid is secreted and chemically breaks down some of the organic compounds into smaller pieces. Proteins are partially digested by pepsin (a protein digesting enzyme).

C. *Small Intestine*.

The pancreas secretes sodium bicarbonate into the upper portion (duodenum) of the small intestine where it neutralizes the acid food entering from the stomach. The pancreas also adds protein, lipid, and carbohydrate digesting enzymes.

The gall bladder discharges bile into the duodenum. The bile helps to emulsify (disperse in water) lipids. This is extremely important since the lipid digesting enzymes are soluble in water and need to come in contact with the lipid molecule in order to digest it.

Cells of the duodenum also produce protein and carbohydrate digesting enzymes. The mucosa of the small intestine absorbs the digested food:

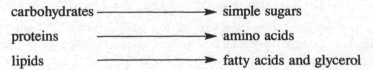

Food consists of complex organic compounds, which can be classified as carbohydrates, lipids, and proteins. The carbohydrates we eat that need digesting are the complex sugars and starch. Usually, only simple sugars (monosaccharides) can be used directly by the cells. Thus, complex sugars and starch must be digested into simple sugars, such as glucose ($C_6H_{12}O_6$). All of the complex sugars and starch are made by forming a chemical bond between simple sugars. This is accomplished by enzymes and results in the removal of a molecule of water (dehydration) each time a bond is formed.

Digestion reverses this process by enzymatically adding a molecule of water (hydrolysis) for each bond broken until digestion is completed and simple sugars remain.

Proteins consist of long chains of amino acids. The amino acids can be used directly by the cells, but the ingested proteins cannot. Again, as in the carbohydrates, the amino acids are chemically bonded together with a molecule of water being lost. Thus, to reverse the process (digest the protein) a molecule of water must be added by means of an enzyme to break the bond between the amino acids.

The most common lipids found in our diet consist of a backbone of glycerol with three fatty acids attached to it. Each time a fatty acid bonds to the glycerol, a molecule of water is lost.

This process is called _____. Recall that dehydration is also involved in carbohydrate and protein formation. In order to digest a lipid, water again must be enzymatically added at the point where a fatty acid molecule bonds to the glycerol molecule.

The reaction that will be studied is shown below. In the presence of the enzyme amylase, starch, a polysaccharide, is digested to maltose, a reducing sugar.

$$\text{Starch (Amylose)} + \text{Water (H}_2\text{O)} \xrightarrow{\text{Amylase}} \text{Maltose (C}_{12}\text{H}_{22}\text{O}_{11}\text{)}$$

From past experience you should recall that the iodine test is used to demonstrate the presence of starch and the Benedict's test the presence of reducing sugars. We will perform tests too demonstrate that hydrolysis has occurred.

Proceed as follows:

1. Obtain a spot plate and label wells 1–6.

2. To wells 1–4 add 2 drops of a 1% starch solution.

3. To wells 5 and 6 add 2 drops of distilled water.

4. To wells 1 and 2 add 2 drops of a 1% amylase solution.

5. Add one drop of iodine solution (I_2KI) to all six wells.

Wherever iodine came in contact with starch a blue-black color change occurred. Now set the spot plate aside and incubate the reaction while you set up the test for sugar. If hydrolysis occurs the blue-black color will diminish or disappear, and that will be a visible indication that hydrolysis has occurred.

1. Obtain 4 test tubes and label them 1–4. Tube 4 will not receive any enzyme and will serve as the control.

2. To all 4 tubes add 2 ml of a 1% starch solution. (Hint: the plastic pipettes are graduated and there is a 1ml mark just below the bulb).

3. Add 2 drops of a 1% amylase solution to tube 1, wait 10 seconds, then add a dropper full of Benedict's solution and place tube 1 in boiling water.

4. Add 2 drops of a 1% amylase solution to tube 2, wait 30 seconds, then add a dropper full of Benedict's solution and place tube 2 in the boiling water.

5. Add 2 drops of a 1% amylase solution to tube 3, wait 60 seconds, then add a dropper full of Benedict's solution to both tubes 3 and 4 and place them in the boiling water.

After 3 minutes remove the tubes from the boiling water and observe the color of each tube. Blue color indicates no sugar present but green, yellow, orange or red indicate increasing amounts of sugar present.

IF THE BENEDICT'S TEST WAS POSITIVE FOR SUGAR YOU MAY DISCONTINUE THE SPOT PLATE EXPERIMENT WHICH WAS A BACKUP TEST THAT TAKES LONGER TO DEVELOP.

1. In which tube or tubes was sugar demonstrated? _____

2. Did hydrolysis of the starch occur more rapidly in the presence of the enzyme?

   _____

3. What happens to amylase activity in the stomach? _____
   (Hint: refer to Exercise 4: Enzymes)

4. What digestive gland produces buffers and additional amylase to allow you to continue the digestion of starch in the intestines? _____
   (Hint: starch belongs to what category of foods?)

II. Gas Exchange

Aerobic respiration at the cellular level involves the step-by-step breakdown of food molecules into carbon dioxide and water, trapping of some of the released energy in the form of ATP, and the use of oxygen. All living cells respire continuously, thus oxygen and carbon dioxide diffuse in and out of them continuously passing along a concentration gradient across the moist plasma membrane of every cell.

Respiration, or gas exchange, at the organismic level is a reflection of cellular respiration in that it involves obtaining oxygen from the environment and releasing carbon dioxide. Plants, in the process of photosynthesis, make use of much of the carbon dioxide. Animals and other organisms without chlorophyll must release this waste product into the air or water in which they live.

As in the case of cellular respiration, gas exchange between an organism and its environment always takes place by diffusion across moist membranes, but oxygen reaches the individual cells in different ways: sometimes by direct diffusion into the cytoplasm from water or air, sometimes by diffusion into a transport system, such as the blood, which carries the gas in solution. Carbon dioxide reaches the environment by traveling the opposite direction in the same manner.

III. The Five Principal Methods of Gas Exchange in Living Organisms

A. *Diffusion across the cell membrane between environment and cytoplasm.*

In land plants, gases diffuse through openings called stomates and lenticels into the air-filled intercellular spaces, where they dissolve in the film of water always present on the cell walls and enter the cells in solution.

Aquatic plants, lacking a cuticle as well as stomates and lenticels, exchange gases across the membranes of epidermal cells. In sponges, flatworms and other small animals, gases can diffuse through the epithelial cells to reach other cells deeper in the body. Because of the thinness of the plants and the small size of the animals, however, no cell is far removed from the surrounding medium, and the distance to be traveled is short.

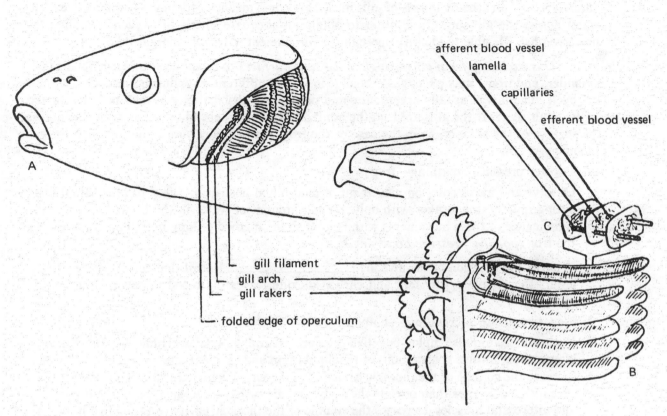

Figure 16–1. A. Gills (of a bony fish) in gill chamber with operculum folded back. B. Part of a gill showing rakers and filaments. C. Part of a filament showing lamella containing the capillaries which aerate the blood. Afferent vessels carry blood to capillaries in gill filaments for oxygenation; efferent vessels collect oxygenated blood and carry it to the dorsal aorta for distribution to head and body.

B. *Diffusion across the skin and other body tissues between environment and blood vessels (skin breathing).* In most higher animals gases are transported by body fluids. This system appears in its simplest form in the earthworm. Oxygen diffuses through the moist body wall into superficial blood vessels, where it is picked up by the blood and transported throughout the body. Carbon dioxide follows this route in reverse.

C. *Tracheae.* In insects and some related animals the circulatory system plays little or no part in gas transport. Instead, these animals have fine, air-filled tubes (tracheae) branching from the body surface to all internal organs and tissues. The external openings of the tracheae are called *spiracles.* Tracheae develop as ingrowths of the body wall and are lined with *chitin,* the wall substance. They end in microscopic intercellular tubules filled with fluid, where gas exchange with adjacent cytoplasm occurs.

Since oxygen diffuses rapidly through air, movement of the body parts alone is sufficient to distribute it throughout the tracheal systems of the small animals which employ this type of respiratory mechanism. Examine the microscope slide labeled "Insect tracheae." If available, a slide of a flea can be examined and tracheae can be seen leading away from the spiracles on the sides of the abdominal segments.

D. *Gills.* Fishes, crustaceans, mollusks and many other aquatic animals breathe by means of gills which are highly vascularized filaments of tissue covered by a thin epidermis. Gases diffuse across the single layer of epithelial cells and across the capillary walls between the surrounding water and the blood, which transports them to and from the cells. Since the concentration of oxygen in water is much lower than in air and its rate of diffusion in water much slower, most animals which breathe by means of gills must have some means of forcing a constant stream of water across the gill surfaces in order to obtain an adequate supply of oxygen. Examine the preserved fish. Be able to identify operculum, gill filament, gill arch and gill rakers.

E. *Lungs. Gas exchange from air through moist lung surfaces to blood vessels.* Amphibians (salamanders and frogs) may effect gas exchange in several different ways during the course of their life cycle. The larvae usually are aquatic and possess gills, which may or may not be retained in the adult stage. If the gills are lost, the adult animals may depend upon gas exchange across the surface of the skin; or, as in the case of the frog, may employ both lungs and skin for gas exchange.

  1. Positive pressure breathing—frog.

     Observe and make notes on movement of mouth and nostrils in a frog or other amphibian sitting quietly in a beaker with only a small amount of water in the bottom. Discuss the significance of these movements with respect to the method of lung breathing employed by the frog *(positive pressure breathing.)*

     What is the function of the alveoli? _____ Recall that this is where oxygen enters the circulatory system and carbon dioxide leaves the blood vessels and enters the alveoli.

  2. Negative pressure breathing—humans.

     Birds and mammals employ negative pressure breathing. The mechanics behind this form of breathing involve contractions of the diaphragm and external intercostal muscles, which increase the volume of the thoracic cavity. This increased volume results in a lower air pressure inside the lungs than outside the body, and air rushes in. Relaxation of the diaphragm and intercostal muscles decreases the volume of the thoracic cavity resulting in a higher air pressure inside the lungs than in the atmosphere and air rushes out of the lungs until the pressures equalize.

     Observe the balloon-bell-jar demonstration of negative pressure breathing, which is employed by humans.

IV. Changes in Dimensions of the Thorax and Abdomen While Breathing (Humans)
(Record data in Table 1).

A. Place a tape measure around your partner's chest at the axillary level (under the arms). Hold the tape measure in place while your partner first breathes normally, then with forced inspiration and expiration. Make measurements under both conditions.

B. Place the tape measure from the lower end of the sternum to the navel and record measurements under conditions of both normal and forced inspiration and expiration.

C. Measure the circumference of the abdomen at the level of the umbilicus (navel) for the same conditions of breathing.

	Normal		Forced	
	Inspiration	Expiration	Inspiration	Expiration
Chest Circumference				
Sternum / Navel				
Abdomen Circumference				

186

Which breathing action (inspiration or expiration) results in an increase in thoracic volume?

_____

During forced inspiration does the circumference of the abdomen increase? _____

Explain how this increase contributes to a greater inspiration of air _____

_____

_____

Under forced conditions, which action (inspiration or expiration) exhibits the greater *change* in overall measurement? _____

Some individuals primarily use their intercostal muscles for breathing. These individuals are termed thoracic breathers. Other individuals also use their intercostal muscles, but rely mainly on their diaphragm and abdominal muscles. These individuals are termed diaphragm or abdominal breathers.

Based on your recorded measurements, which type of breather are you? _____

Which type of breather would be expected to move a greater volume of air in and out of the lungs?

_____

V. Measurement of Lung Capacity

The spirometer will be used in the measurement of the several volumes of air that are included within the capacity of our lungs. The volume of air which enters and leaves the lungs with each normal breath is known as the *tidal air*. The volume which can be taken in over and above the tidal inspiration is called *inspiratory reserve*. The volume that can be expelled after a tidal expiration is called the *expiratory reserve*. The sum of these volumes is equal to the *vital capacity*. The volume of air remaining in the lungs that cannot be expired is the residual air. For an average person the volume of residual air is about 1,200 cubic centimeters. The *total lung capacity* is equal to the sum of the residual volume plus vital capacity.

In the following graph fill in each of the volumes in liters and compute the vital capacity and total lung capacity for each person at your table. (1 liter = 1,000 cubic centimeters).

Directions for Using the Spirometer

A. The volume scale should be on "0".

B. Put a clean mouth piece on the end of the tube (Note: only *exhale* into the tube).

C. After a normal inhalation, exhale into the spirometer tube. The reading on the volume scale indicates *tidal volume.*

D. After a normal inhalation, forcibly exhale as much air as you can into the tube. The reading on the volume scale indicates tidal volume plus expiratory reserve volume. To find your *expiratory reserve volume,* simply subtract the tidal volume value from the reading on the scale.

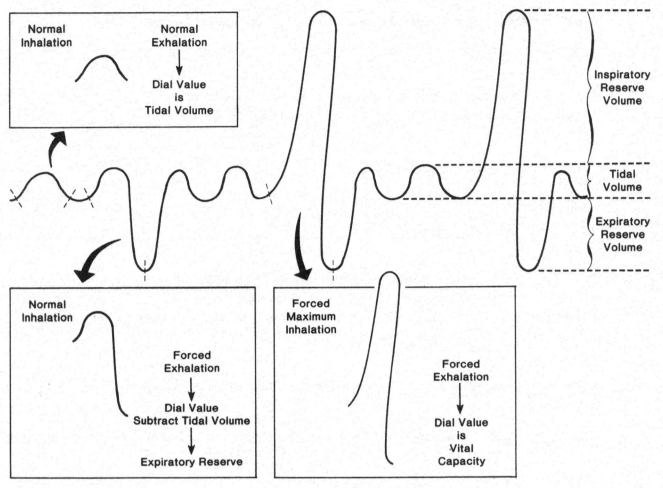

Figure 16-2.

E. Inhale as much air as you can and forcibly exhale into the tube. The reading on the volume scale is your *vital capacity*.

F. To determine your inspiratory reserve volume, subtract both your tidal volume and expiratory reserve volume from your vital capacity. The remainder is your *inspiratory reserve volume*.

VI. Chemistry of Breathing

Breathing is controlled by a "breathing center" in the medulla of the brain. This center is activated by a buildup of carbon dioxide in the blood and/or a low oxygen concentration, such as occurs at high elevations.

To determine variations in respiratory rate have your partner count the number of your inhalations per minute while you are sitting quietly. Then have him record the same data after you have run in place for one and one-half minutes. *Why is there a difference?*

VII. Aerobic Respiration

The rate of aerobic respiration can be determined by measuring the amount of glucose or oxygen utilized, or by measuring the amount of water or carbon dioxide produced. In this exercise, carbon dioxide production will be measured.

A. Measuring Carbon Dioxide Production

1. Prepare a 0.04% solution of sodium hydroxide (NaOH) from a 0.4% stock solution of NaOH by adding 90 ml of distilled water to 10 ml of stock solution. Place the 0.04% NaOH solution in a burette tube.

2. Obtain a solution of phenolphthalein for use in measuring carbon dioxide production. Phenolphthalein is an indicator which turns pink at an alkaline pH such as after all the carbon dioxide has been removed from the water sample.

3. The method being used to measure carbon dioxide production is based on the following chemical reactions.

   a. Carbon dioxide combines with water to form carbonic acid when you blow air from your lungs into the water.

$$CO_2 + H_2O \rightarrow H_2CO_3$$

   b. Sodium hydroxide when added combines with carbonic acid to form sodium bicarbonate.

$$NaOH + H_2CO_3 \rightarrow NaHCO_3 + H_2O$$

   c. When a measured amount of alkali (NaOH) is added, it reacts with the acid until it is neutralized. The addition of any more alkali after this point has been reached will change the solution's color. Since equal amounts of equal concentrations of acid and alkali react, the amount of sodium hydroxide added will equal the amount of carbonic acid in the water.

B. Measuring Carbon Dioxide Production in Humans

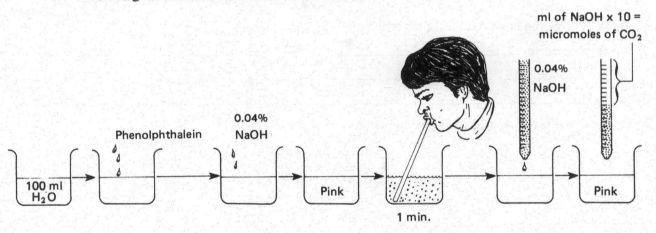

Place 100 ml of tap water in a bottle or beaker, and add three to five drops of phenolphthalein indicator. If, after adding the indicator, the water is not pink add 0.04% NaOH, drop by drop, until the pink color appears. Then, bubble air from your lungs into the water through a soda straw for exactly one minute. What happens to the color of the indicator as the carbon dioxide from your breath combines with water to form carbonic acid?

A change in the hydrogen ion content (pH) of a solution causes rearrangement of the indicator molecule resulting in a color change. Phenolphthalein becomes colorless.

After the one minute, add 0.04% sodium hydroxide solution slowly from a burette tube. Add enough NaOH to make the solution turn pink and stay pink for 30 seconds after shaking. Observe and record the ml of 0.04% NaOH which you added. Multiply this number by 10. The product is the number of micromoles of carbon dioxide which you exhaled in one minute.

189

C. Calculating Respiration Rate

R (number of micromoles of CO_2 per gram of organism per hour), the rate of respiration can be calculated as follows:

$$R = \frac{N}{W \times T}$$

Where N is equal to the number of micromoles of CO_2 exhaled in one minute, W is equal to your weight in grams, and T equals time in hours.

Note: One pound equals approximately 454 grams.

D. Correlating Body Weight with Respiration Rate.

Collect class data of rate of respiration and body weight and record on graph below. Is there a correlation between body weight and rate of respiration? Discuss your answer.

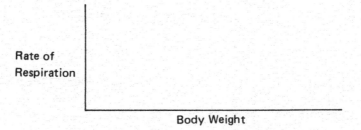

Rate of
Respiration

Body Weight

REVIEW QUESTIONS

1. What is (are) the end products of digestion of:

   a. carbohydrates _____

   b. proteins _____

   c. lipids _____

2. Define dehydration and hydrolysis reactions.

3. Define digestion.

4. List the main digestive functions which occur in each of the following:

   a. mouth cavity

b. stomach

c. small intestine

5. List the five principal methods of gas exchange and identify at least one organism that uses each method.

Method	Organism
a. _____	_____
b. _____	_____
c. _____	_____
d. _____	_____
e. _____	_____

6. Define:

   a. vital capacity:

   b. tidal air:

   c. residual volume:

7. Contrast aerobic respiration with gas exchange.

8. What function does the phenolphthalein serve?

9. What waste product of respiration was measured to indicate the rate of cellular respiration?

10. What was the function of adding NaOH into the beaker *before* bubbling in carbon dioxide?

VERTEBRATE ANATOMY:
THE CIRCULATORY SYSTEM AND BLOOD

Objectives

Upon completion of this exercise you should be able to:

1. List a function for each of the main types of blood cells and the platelets.
2. Prepare and stain a blood smear.
3. Identify, microscopically, erythrocytes and leucocytes.
4. Prepare and interpret a graph or histogram showing the distribution of blood clotting times for a group of individuals.
5. Identify the following parts from a heart specimen, model or diagram: anterior and posterior vena cavae, right atrium, tricuspid valve, right ventricle, semilunar valves, pulmonary artery, pulmonary veins, left atrium, bicuspid or mitral valve, left ventricle, aorta and interventricular septum.
6. Contrast arteries, veins and capillaries.
7. Distinguish an artery from a vein microscopically.
8. Contrast open and closed circulatory systems.
9. Identify the following blood vessels from a fetal pig or fetal diagram: pulmonary artery, aorta, renal arteries, renal veins, anterior vena cava, posterior vena cava, umbilical arteries, umbilical vein and ductus arteriosus, and others as noted by your instructor.
10. Answer all review questions in the exercise.

I. Introduction

As animals increased in size, their cells became more distant from the external environment. Complex transport systems evolved for moving oxygen and various nutrients to cells and for picking up carbon dioxide and other wastes from cells rapidly enough to maintain life. Vertebrates evolved a circulatory system in which a fluid, *blood,* is pumped by a muscular *heart* through a continuous network of *blood vessels.* In this exercise you will examine selected aspects of blood, of the heart and of the blood vessels.

II. Blood

Blood is a complex substance consisting of *cells* and a liquid, *plasma.* The plasma is mostly water with a number of inorganic and organic substances dissolved in it. Among the organic substances are several types of blood proteins which help maintain proper water concentration, aid in blood clotting and help the body combat disease organisms. These latter, called antibodies, frequently work in conjunction with phagocytic cells. All substances used in cells, including amino acids, simple sugars and fats, as well as cell produced wastes such as carbon dioxide and urea, are found in the plasma.

Note: Caution should be used in working with blood and all blood contaminated items should be placed in the designated disposal container when finished.

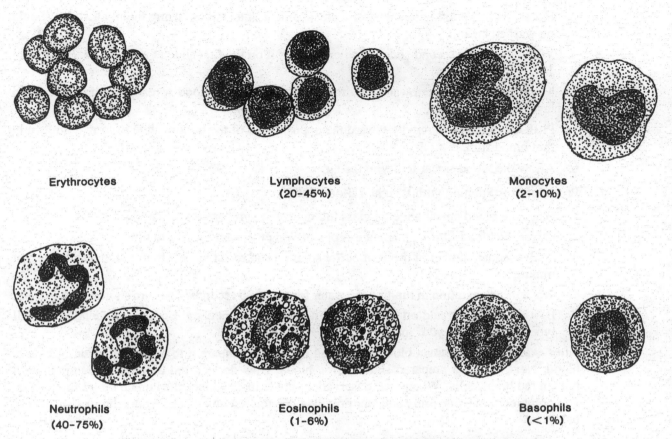

Erythrocytes

Lymphocytes
(20-45%)

Monocytes
(2-10%)

Neutrophils
(40-75%)

Eosinophils
(1-6%)

Basophils
(<1%)

Figure 17-1. Types of human blood cells and relative frequencies of leucocytes.

A. Blood Cells and Platelets (Figure 17-1).

The kinds of cells in human blood are *erythrocytes* (red blood cells) and *leucocytes* (white blood cells). In a given volume of blood the nonnucleated erythrocytes are the most numerous and appear biconcave and pink in color. The red pigment hemoglobin is localized in them and functions in transporting oxygen and carbon dioxide. Leucocytes are much larger than the erythrocytes. Some leucocytes are amoeboid and protect the body against disease and infection by engulfing bacteria and other foreign materials. These are *phagocytes* and the process is called *phagocytosis*. Phagocytes are frequently assisted in this process by *antibodies* produced by other leucocytes.

Leucocytes may be divided into three groups with respect to their nuclear characteristics:

1. *Lymphocytes:* 7-14 μm in diameter; nucleus spherical.

2. *Monocytes:* 12-18 μm in diameter; nucleus eccentric in position and oval to bean-like in shape.

3. *Polymorphonuclear leucocytes* (= Granulocytes): 10-14 μm in diameter; nucleus variable in shape; frequently lobed. Granules are abundant in the cytoplasm; three types, *basophils, eosinophils,* and *neutrophils,* are distinguishable based on the staining reaction of their cytoplasmic granules to Wright's stain. Granules in the basophils stain deep blue, in the eosinophils bright red, and in the neutrophils lilac.

Platelets are bits of cytoplasm that are much smaller than red blood cells. They contain an enzyme that plays a key role in the clotting of blood. To observe human blood cells prepare a blood smear and stain it with Wright's stain.

B. Steps in the Preparation of a Blood Smear.

1. Clean an index finger using sterile cotton moistened in 70% alcohol, let it air dry and prick it with a sterile disposable lancet.

193

2. Put a drop of blood on one end of a clean slide. (Note: Before using them, all slides should be washed clean.)

3. While holding a second slide, place one end of it on the first slide at a 30° angle. See Figure 17-2.

4. Move the upper slide toward the drop of blood until the blood spreads forming a uniform layer along the edge.

5. Push the upper slide rapidly toward the opposite end of the bottom slide to form a uniformly thin blood film.

6. Allow the blood smear to dry in the air.

C. Procedure for Staining with Wright's Blood Stain.

1. Place the blood smear slide on glass tubing over a syracuse dish. See Figure 17-3.

2. Cover the blood film with about 8 drops of Wright's stain for one to three minutes.

3. Add an equal number of drops of buffer solution and rock the slide to mix. Let stand for eight minutes.

4. Wash the preparation in running tap water for about 30 seconds. The film should be pale pink.

5. Blot the slide dry, add oil and examine it under oil immersion. Locate and identify the cell types indicated by your instructor.

After examining the human blood smear you prepared, compare it to a prepared slide of human blood. Next, obtain a prepared slide of frog blood. How do frog red blood cells compare with human red blood cells? When your observations are complete place the human blood smear slide you prepared in the disposal container provided. Sketch red blood cells from a human and a frog below.

D. Blood Clotting.

At the same time you obtain blood for preparing a smear draw blood into a capillary tube. Direct the tube downward so that gravity will assist in filling it. To observe clotting, carefully break the capillary tube after four to five minutes, drawing the broken part away to observe the threadlike fibrin.

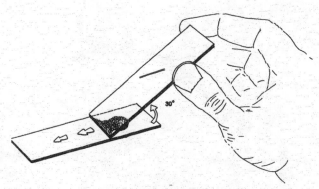

Figure 17-2.

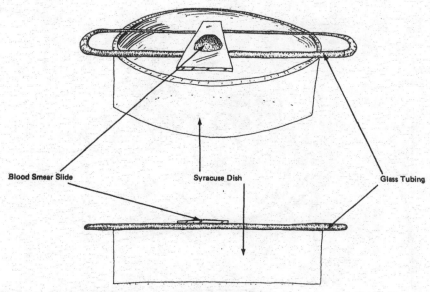

Blood Smear Slide   Syracuse Dish   Glass Tubing

Figure 17-3.

III. Heart (Fig. 17-4)

Obtain a sheep heart for observation. A larger beef heart may also be in the laboratory for observation. Place the heart in front of you in a dissecting pan with the posterior, pointed end, or apex, toward you. The right side of the heart will be on your left. The mammalian heart is enclosed in a thin membranous sac called the *pericardium*. Yellowish fat deposits may also be seen on the outside of the *myocardium* or heart muscle. The inner compartments of the heart are lined by another thin membrane, the *endocardium*.

There are four compartments or chambers. The anterior chambers called *atria* receive blood from the body. The posterior chambers called *ventricles* pump blood out to all parts of the body. Now make an incision from anterior to posterior through the left atrium and left ventricle. Make a similar incision through the right atrium and right ventricle. Which ventricle has the thickest wall? _____ Why? _____

Obviously the left ventricle has the thicker wall because it pumps blood out to all parts of the body. By now you should be able to see that the heart is essentially a double pump. Each pump, separated by a wall or *septum* from the other, consists of an anteriorly located atrium, a one-way valve and a posteriorly directed ventricle. Blood returning from the posterior body including the legs and abdominal organs enters the *right atrium* via the *posterior vena cava* (inferior vena cava in humans). Blood returning from the anterior body including the head and neck enters the right atrium via the

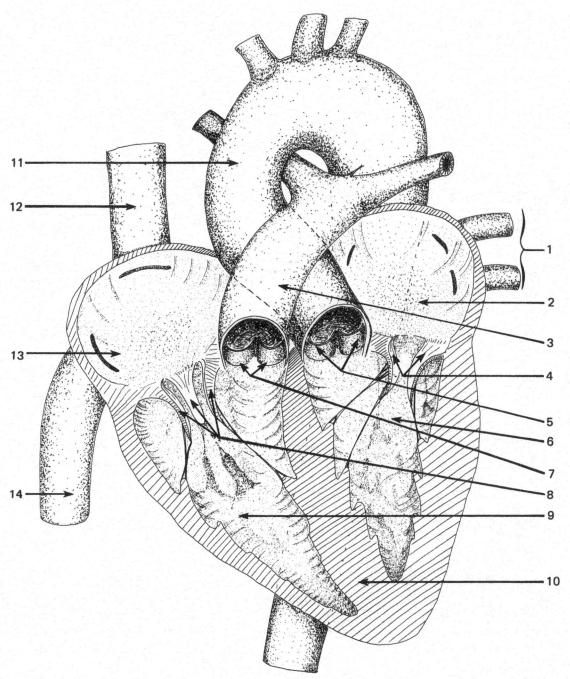

Figure 17-4. The human heart.

anterior vena cava (superior vena cava in humans). Both atria contract simultaneously forcing blood into the ventricles. Blood leaving the right atrium passes through a one-way valve, the *tricuspid valve*, into the *right ventricle*. Then, both ventricles contract simultaneously, forcing blood out of the heart.

Fibrous cords, connected to the one-way valves, can be seen in the ventricles. These prevent the valves from flapping up into the atria when the ventricles contract. Blood leaving the right ventricle passes through a *semilunar valve* into the *pulmonary artery* which branches, carrying blood to both lungs for gas exchange. Blood returns from the lungs via *pulmonary veins* to enter the *left atrium*. Blood leaving the left atrium passes through the *bicuspid* or *mitral valve* into the *left ventricle*. It is then pumped out, passing through a *semilunar valve* into the *aorta*. Branches from the aorta transport blood to all parts of the body.

Using the heart as a reference point, a drop of blood circumscribes two circuits in one complete trip through the body. One circuit, the *pulmonary,* carries blood from the heart to the lungs and back to the heart. The other, the *systemic,* carries blood from the heart to the body and back to the heart. Thus blood flows from heart to lungs to heart to body to heart, etc.

After finding the italicized parts above on the sheep heart, locate and label the same parts of the human heart (Figure 17–4).

IV. Blood Vessels

In the closed circulatory system of mammals blood flows through a continuous network of blood vessels. Blood leaves the heart under pressure to enter thick-walled, elastic *arteries.* Blood returns to the heart in thinner walled *veins.* In between, the smallest arteries and veins are connected by *capillaries.* The wall of a capillary is one cell layer in thickness. It is at the capillary that the exchange of materials occurs between the blood and the cells.

Obtain a prepared slide on which are cross sections of an artery and a vein. Arteries do not appear collapsed in cross section. They have a relatively thick wall and a small lumen. The wall consists of three coats: an inner lining of endothelium; a middle coat consisting of smooth muscle, elastic and white fibrous connective tissue; and an outer coat of white fibrous connective tissue. The middle coat permits dilation or constriction of arteries. Veins have the same three coats. However, the wall of a vein appears collapsed in cross section, is thinner with fewer elastic fibers and the lumen is relatively larger.

The wall of capillaries consists of a single layer of cells, the endothelium, which is continuous with the endothelium of arteries and veins. A capillary may be so narrow in diameter that red blood cells must pass through in single file.

Contrast the structure of an artery and vein in terms of the function of each type of vessel.

Large arteries can be seen to pulsate, that is, expand and recoil. The variation in flow is due to the beating of the heart. As the heart contracts strongly in *systole,* blood is pushed into the arterial system under considerable pressure. The elastic walls of the arteries expand in response to the sharp increase in blood pressure. As the heart relaxes and fills, in *diastole,* pressure in the arteries decreases. A steadier flow under decreased pressure is observed in the veins.

In an open circulatory system the flow of blood is slow and sufficient for small organisms. It does not flow through a continuous network of vessels. Instead, it is pumped from open-ended vessels into body spaces called *hemocoels* where blood bathes the cells directly. Open transport systems are common only among the invertebrates, where the animals are small and often have supplementary means of transporting oxygen and nutrients to cells, and carrying wastes away from them.

V. Fetal Pig Circulatory System

Obtain the pig you used previously and place it in a dissecting pan as before. The major blood vessels, arteries and veins are injected with latex to facilitate observation. Supposedly, the arteries are filled with red latex, the veins with blue. This may not always be the case because the pressure of injection may have forced blood and latex through heart valves against the usual flow pattern and may have even crossed capillary beds in some places.

A. Arterial System (Fig. 17–5).

Find the heart in the thoracic cavity of the pig. Identify the pericardium; posterior, thick-walled ventricles; and anterior, thin-walled atria. Locate the pulmonary artery. It is the large vessel which carries blood from the right ventricle of the heart. Trace the *pulmonary artery* until it branches by rolling the heart to one side. The right branch goes to the right lung, the left branch to the left lung. Since the lungs aren't functional during fetal existence, most of the blood in the pulmonary artery bypasses the lungs and goes to the aorta through the *ductus arteriosus.* Find it. (The ductus arteriosus is not visible in the adult heart because it is obliterated soon after

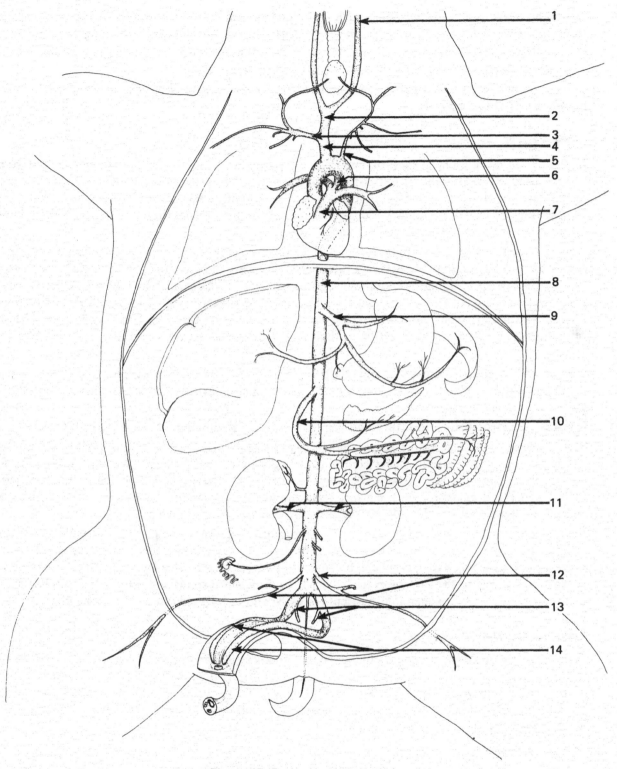

Figure 17-5. Arteries of a fetal pig.

1

2
3
4
5
6

7

8

9

10

11

12

13

14

birth.) Next, locate the large, thick-walled *aorta* that carries blood from the left ventricle. It passes anteriorly for a short distance and then makes a U turn to go posteriorly along the dorsal wall of the thoracic and abdominal cavities. Starting with the heart and proceeding posteriorly, identify each vessel called for which branches from the aorta.

First locate two branches that emerge from the *aortic arch* a short distance from the heart. The first is the *brachiocephalic artery,* which supplies blood to the right forelimb. This branch continues anteriorly, then divides into two branches. One, the right subclavian artery, supplies blood to the right forelimb. The second branch, the *bicarotid trunk,* continues anteriorly, then divides into two *common carotid arteries.* The second branch off of the *aortic arch* is the *left subclavian artery.* It carries blood to the left forelimb.

The first major branch of the *aorta* after entering the abdominal region is the *celiac artery.* Locate it in the region dorsal to the stomach. The celiac artery subdivides many times to send branches to the stomach, spleen and liver. The *anterior mesenteric artery* branches from the aorta a short distance posterior to the celiac artery. It sends branches to the pancreas, small and large intestines. Find it by picking away the mesenteries with a probe. Continue posteriorly and find the paired *renal arteries* leading to the kidneys. A short distance posterior to the kidneys the aorta divides into two *external iliac arteries* that lead to the hind legs. The large *umbilical arteries* pass ventrally from the internal *iliac arteries* at about this point to go to the umbilical cord. The paired internal iliac arteries are located posterior to the paired external iliac arteries.

B. Venous System (Figure 17-6).

Find the large vein that enters the right atrium anteriorly. This is the *anterior vena cava.* Several pairs of veins that drain the anterior region of the body unite to form this vessel. Locate the following veins beginning at the midline and proceeding to your right: 1. the *internal jugular vein* from the brain; 2. the *external jugular vein* from the head and neck—the internal and external jugular veins parallel each other; 3. the *subscapular vein* from the shoulder muscles; 4. the axillary vein and its continuation, the brachial vein, from the leg; 5. the subscapular vein and the axillary vein merge with the jugular veins forming the short *brachiocephalic* (innominate) vein that enters the vena cave (note: the subscapular and axillary veins often join to form a subclavian vein before entering the brachiocephalic vein); 6. the *costocervical vein* from muscles of the neck and back.

Now, locate the large *posterior vena cava.* It can be found paralleling the aorta in the lower abdomen. Find the paired *renal veins* that join the posterior vena cava at the kidney level. Next, locate the *umbilical vein* from the placenta. This may have been cut when you opened the abdominal region. The umbilical vein is a fetal structure that closes after birth.

In the adult, arteries supply the abdominal organs with blood that passes through the capillaries of the digestive tube, enters the liver via the *portal vein* and again passes through a system of capillaries (the sinusoids of the liver) before entering the posterior vena cave through the *hepatic veins.*

The posterior vena cave continues anteriorly to join the right atrium.

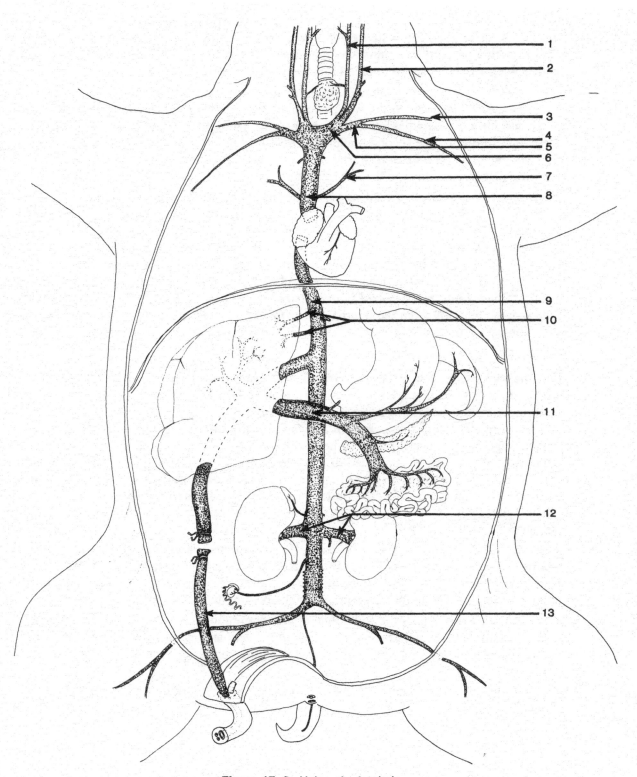

Figure 17-6. Veins of a fetal pig.

REVIEW QUESTIONS

The principal components of the vertebrate circulatory system are:

1. A fluid, _____ .

2. A muscular _____ for pumping; and

3. A continuous network of _____ .

4. The vertebrate circulatory system is the (open, closed) type.

   The kinds of formed elements in the blood are:

5. _____ , which transport oxygen and carbon dioxide;

6. _____ , which function in protection against disease; and

7. _____ , which aid in the blood clotting reactions.

8. Blood vessels carrying blood away from the heart are called _____ .

9. Thin-walled vessels that connect arteries to veins are called _____ .

   Blood enters the right atrium via two major veins:

10. The _____ and the

11. _____ .

12. Blood drops from the right atrium into the right ventricle via the _____ valve.

13. Blood is pumped from the right ventricle through a semilunar valve into the _____ .

14. Gas exchange occurs in the lungs and the blood returns to the left atrium of the heart via the

    _____ .

15. Blood drops from the left atrium via the _____ valve into the left ventricle.

16. Blood is then pumped out through a semilunar valve into the _____ which branches to distribute blood to all parts of the body.

17. The _____ circuit carries blood from the heart to the lungs for oxygenation and back to the heart.

18. The _____ circuit carries blood to the body except the lungs for oxygenation and back to the heart.

Exercise 18

VERTEBRATE ANATOMY: THE URINARY AND REPRODUCTIVE SYSTEMS

Objectives

Upon completion of this exercise you should be able to:

1. Determine the sex of a fetal pig.
2. Identify the following parts of the urinary system of the pig: kidney, ureter, renal hilus, urinary bladder, urethra, urogenital sinus.
3. Identify the following on models: kidney, ureter, urinary bladder, urethra, nephron, glomerulus, Bowman's capsule, tubule, collecting duct, renal pelvis.
4. Recognize from a microscope slide renal cortex and medulla, glomerulus, Bowman's capsule and tubules.
5. Locate the following parts of the urogenital system on a fetal pig or a diagram of a fetal pig: urethra, urogenital sinus, ovaries, oviducts (uterine tubes), horns of the uterus, body of the uterus, vagina, female urogenital opening, urogenital papilla, male urogenital opening, penis, vas deferens, epididymis, testes, scrotal sacs, seminal vesicles, bulbourethral (Cowper's) glands.
6. Trace the path of sperm from the seminiferous tubules to the urethra of man using models and/or diagrams. Include the glands which form the seminal fluid.
7. Trace the paths of sperm and ovum to the point of fertilization in the uterine tube of the female using models and/or diagrams.
8. Describe the male and female reproductive systems.
9. Give a function for each of the structures listed in objectives 2, 3 and 5.
10. Answer all review questions and label all figures in the exercise.

I. Introduction

The urinary system is the principal system for the elimination of nitrogenous wastes from the body. It also plays an important role in maintaining ionic and fluid balance of the blood, pH of the blood and blood pressure. In this exercise you will dissect the major organs of the fetal pig urinary system and you will observe features of the human kidney from models, microscope slides, and diagrams. The anatomy of the reproductive system will also be studied through pig dissection, and the study of models of the human pelvis.

II. Fetal Pig Sex

Determine the sex of your pig if you have not already done so. If a small opening is present just posterior to the umbilical cord, then your pig is a male; this is the urogenital opening of the penis. If the urogenital opening is located ventral to the anus, and if there is a urogenital papilla present, you have a female. Since there are sexual differences in the anatomy of the urinary and reproductive systems be sure to look at pigs of both sexes.

Locate the paired *kidneys* lying against the dorsal wall of the abdominal cavity. Notice that each kidney is enveloped by a thin membrane, the peritoneum, on its ventral side. Next, locate the *adrenal gland*, a narrow whitish body about one cm long, along the anteromedial portion of the kidney. If you have not already done so, remove the peritoneum from a kidney and adjacent area to reveal structures entering and leaving a depression, the *renal hilus,* on the medial side of the kidney. You have already examined the paired renal arteries and veins in a previous exercise. The hilus is also the point where the *ureter* leaves the kidney carrying urine to the *bladder.* Trace one of the ureters posteriorly until it enters the bladder. During fetal life urine exits from the bladder anteriorly by way of the *allantoic duct* through the umbilical cord to the placenta.

In the adult male the bladder is drained by the *urethra* which passes posteriorly for a short distance then turns anteriorly and ventrally to enter the penis.

Locate the urethra as it leaves the urinary bladder. With a razor blade or scalpel split the skin on the midline between the legs of the pig. Cut through the muscle and the soft pelvic bones. Remove some of the tissues and locate the urethra below. Observe the change of direction as the urethra turns anteriorly and then ventrally to enter the penis. The urogenital opening is just posterior to the umbilicus. In a female pig also the skin, muscle and pelvic bones must be cut to observe the course of the urethra.

In the adult female the urethra continues posteriorly to enter the *urogenital sinus* approximately one cm from the urogenital opening. Locate all the italicized structures in this section before completing the exercise.

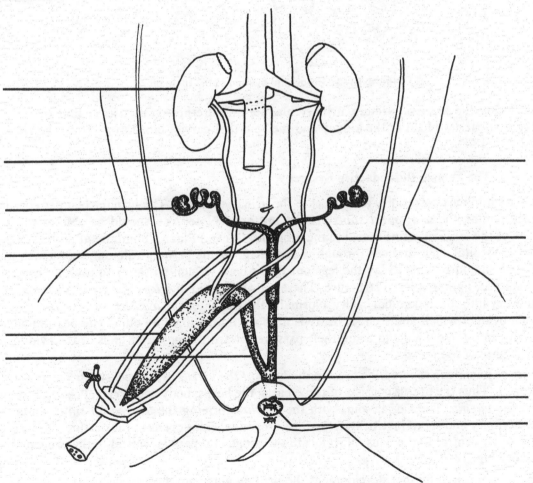

Figure 18–1. Urogenital system of the female pig.

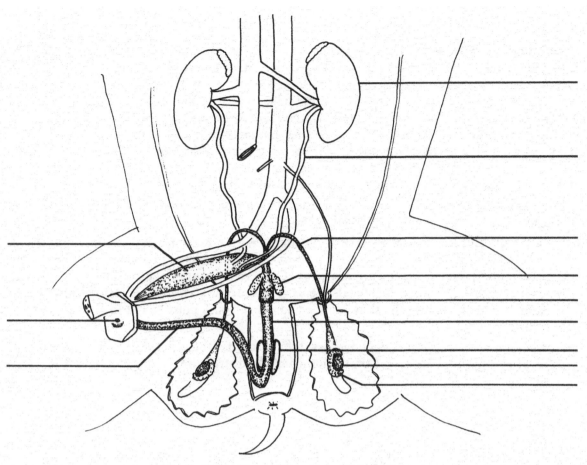

Figure 18–2. Urogenital system of the male pig.

Next, locate the kidneys, ureters, urinary bladder and urethra on models of humans. Identify the artery that carries blood to the kidney and the vein that carries blood away from the kidney. Label Fig. 18–3

IV. Fetal Pig: The Female Reproductive System

Pull the umbilicus posteriorly and relocate the urinary bladder. The urinary bladder continues posteriorly joining the *urogenital sinus* via the *urethra*. (Note: You should be able to see a straight portion of the large intestine dorsal to the urogenital sinus.) Place a blunt metal probe along one side of the urogenital sinus and cut through the cartilage of the pelvic girdle until you reach the probe. Now you should be able to lay the legs out flat. About ½ inch posterior to each kidney is a small, bean-shaped, lighter colored structure. These are the paired *ovaries*. Mature ova are released from each ovary into the abdominal cavity. Normally the ova enter the *oviducts* or uterine tubes. The end of each oviduct which abuts on an ovary is an open funnel lined by cilia. The beating cilia tend to draw ova into the oviducts to begin their passage toward the posterior ends of the oviducts, the so-called *horns of the uterus.*

The horns join along the midline to form the *body of the uterus,* which constricts posteriorly forming the cervix. Thus the uterus consists of three parts: the horns, body and cervix. Posterior to the uterus locate the *vagina.* The vagina joins with the *urethra* forming the *urogenital sinus,* an area shared by the urinary and reproductive systems. The urogenital sinus opens to the outside via the *urogenital opening* just ventral to the anus. Now locate the urogenital papilla ventral to the urogenital opening and label Fig. 18–1.

In humans through further development the urinary and reproductive systems of the adult female acquire separate openings to the outside. Are separate openings for the urinary and reproductive systems present in the adult male?

204

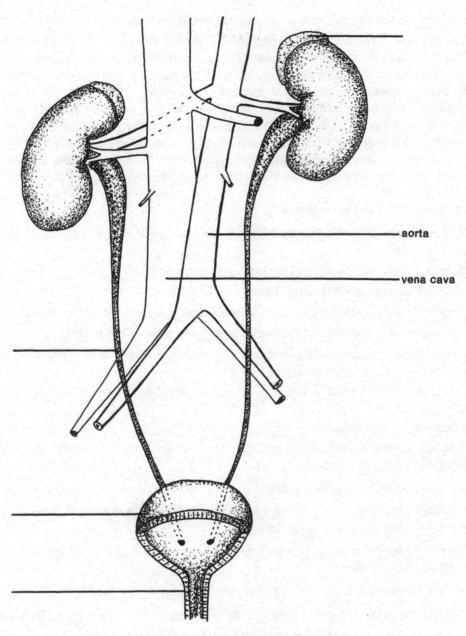

aorta

vena cava

Figure 18-3. The human urinary system.

V. Fetal Pig: The Male Reproductive System

Pull the umbilicus posteriorly and locate the urinary bladder. Just behind the posterior end of the bladder locate two small tubes, each of which loops over a ureter and an umbilical artery and joins medially to the urethra near where it emerges from the bladder. These are the *vasa deferentia* (singular: vas deferens). Blood vessels and nerves parallel each vas deferens and together with it constitute the spermatic cord.

Cut through the skin and trace a vas deferens posteriorly until it joins to a highly coiled mass of tubules, the *epididymis,* which originates at a *testis.* The oval testis is enveloped by a thin membrane. Remove it. Sperm are produced in the paired testes and stored in the paired epididymi until released during copulation. The testes begin development in the abdominal region, but gradually migrate posteriorly into paired *scrotal sacs* located between the hind legs. The path the testes follow as they move toward the scrotal sacs becomes the *inguinal canal* which opens into the abdominal region via

205

the *inguinal ring.* Each vas deferens passes through an inguinal ring to enter the abdominal region. Near where the vasa deferentia join the urethra and dorsal to the urethra locate a pair of small, light colored, lobulated glands, the *seminal vesicles.* A small rounded gland, the *prostate,* is also located in this area. Next, trace the urethra posteriorly until it makes a U-turn and proceeds anteriorly until it opens to the outside via the urogenital opening of the penis. (Note: You will need to cut through the cartilage of the pelvic girdle on one side.) Now locate a pair of white glands, *bulbourethral glands,* one on each side of the urethra near the U-turn. The three types of glands form a fluid, *seminal fluid,* which together with *sperm* constitute the *semen.* By now you should have observed the semen and urine have a common pathway to the outside in the male from the point at which the vasa deferentia join the urethra. Label Fig. 18–2 before proceeding to the next section.

VI. Human Reproduction and Development

1. The male reproductive system (Fig. 18–4): identify the structures listed below on models and/or diagrams.

   Scrotum—the sac in which the testes are located.

   Testes—the sperm producing organs of the male.

   Epididymis—the coiled tubule in which sperm are stored prior to ejaculation.

   Vas deferens—the tube which transports sperm toward the urethra.

   Seminal vesicles, prostate gland, bulbourethral glands—three types of glands which, collectively, form the seminal fluid.

   Urethra—a tube which transports urine from the bladder to the outside and semen to the outside.

   Penis—the male copulatory organ.

2. The female reproductive system (Fig. 18–5): identify the italicized structures listed below on models and/or diagrams.

   Ovary—the ovum producing organ of the female.

   Uterine tube (oviduct)—tube in which ova are transported toward the uterus. Ova are fertilized as they move down the tube.

   Uterus—muscular organ in which the embryo embeds and in which embryonic and fetal development occurs.

   Cervix—constricted portion of the uterus adjacent to the vagina.

   Vagina—tubular organ in which sperm are deposited; also serves as the birth canal.

   Clitoris—the female equivalent of the penis in the male.

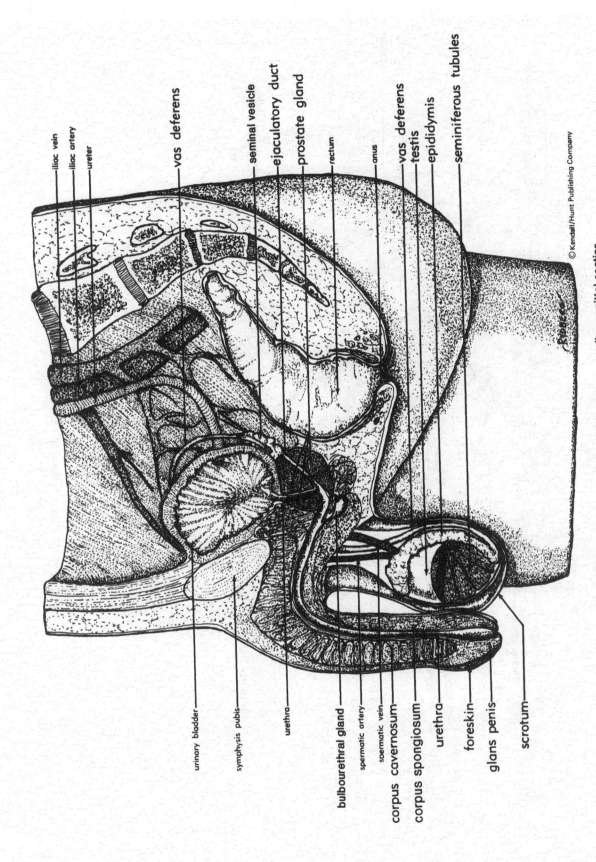

iliac vein

iliac artery

ureter

vas deferens

seminal vesicle

ejaculatory duct

prostate gland

rectum

anus

vas deferens

testis

epididymis

seminiferous tubules

urinary bladder

symphysis pubis

urethra

bulbourethral gland

spermatic artery

spermatic vein

corpus cavernosum

corpus spongiosum

urethra

foreskin

glans penis

scrotum

© Kendall/Hunt Publishing Company

Figure 18–4. Human, male reproductive system, median sagittal section.

207

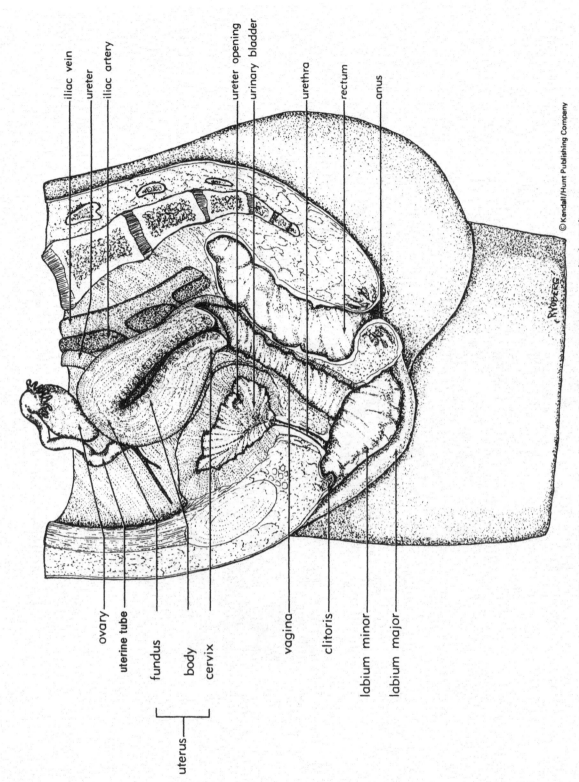

iliac vein

ureter

iliac artery

ureter opening

urinary bladder

urethra

rectum

anus

ovary

uterine tube

fundus

body

cervix

uterus

vagina

clitoris

labium minor

labium major

RYBERG

Figure 18–5. Human, female reproductive system, median sagittal section.

VII. Internal Structure of the Kidney

Examine Fig 18–6 and/or a model of a kidney. Label the *renal artery, renal vein* and *ureter*. The interior of the kidney is composed of:

1. an outer region called the *renal cortex.*
2. an inner region called the *renal medulla,* and
3. a central funnel shaped space called the *renal pelvis.*

Label these three regions on Fig 18–6.

Note that the medulla region consists of many pyramid shaped structures. Urine drains out of the tip of each pyramid into a small channel (minor calyx). Several minor calyces (plural) merge into a larger channel (major calyx) and several major calyces drain into the renal pelvis.

The functional units of the kidney are called nephrons. Most of each nephron is located in the cortex region. Each kidney consists of about a million nephrons. Examine a model showing the nephron, identify the italicized structures described below and label Fig. 18–7. As you read the following description, fill in the missing words based on what you have already learned in this exercise. Each nephron consists of a renal corpuscle and a tubule which is surrounded by an intricate capillary system. The renal corpuscle is composed of a tuft of capillaries called the *glomerulus* and a cup-shaped depression called *Bowman's capsule.* Notice that Bowman's capsule closely envelopes the glomerulus. The tubule portion of the nephron extends from the Bowman's capsule and is divided into three regions. The highly coiled portion closest to the Bowman's capsule is called the *proximal convoluted tubule.* The tubule next straightens out and dips down into the medulla of the kidney as the *loop of Henle,* consisting of a thin descending limb, the U-shaped turn and finally the ascending limb which brings the tubule back into the renal cortex. In the renal cortex, the tubule once again becomes coiled as the *distal convoluted tubule.* The distal convoluted tubule empties into a *collecting duct* which receives urine from many nephrons and carries the urine toward the renal pelvis. Note the network of capillaries, the *peritubular capillaries,* which wrap around the proximal and distal convoluted tubules. They are continuous with a network of capillaries, the *vasa recta,* which follows the loop of Henle.

The blood supply entering the kidney from the renal artery eventually leads to the glomerulus.

A filtrate similar in composition to blood plasma passes from the glomerulus into _____'s capsule. The fluid leaving Bowman's capsule is modified by the selective reabsorption of materials from the tubule into the surrounding capillaries and by secretion of materials into the tubule from the capillaries. This changed fluid, now called urine, leaves the nephron and flows through

the _____ duct toward the renal _____ . Finally the urine trickles down the

_____ to the urinary bladder where it is stored until voided.

VIII. Microscopic Anatomy of the Kidney

Examine a prepared slide labeled Mammalian kidney c.s. with the scanning objective. The renal cortex can be identified by the presence of many renal corpuscles. The renal corpuscles are small round structures, surrounded by a clear space. The renal medulla does not contain renal corpuscles.

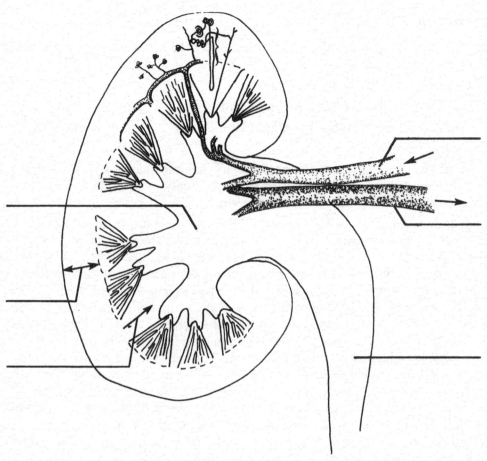

Figure 18-6. The kidney.

Now that you have distinguished between the renal cortex and the renal medulla, place the pointer on the curved outer surface of the kidney and switch to the low power objective. Identify the thin fibrous capsule that covers the kidney. Next move the mechanical stage until you find a renal corpuscle. Center the corpuscle and switch to the high power objective. The center portion of the corpuscle is occupied by a clump of capillaries called the _____ . The clear space surrounding the glomerulus is the cavity (lumen) of _____ capsule. Draw a renal corpuscle and label glomerulus and Bowman's capsule.

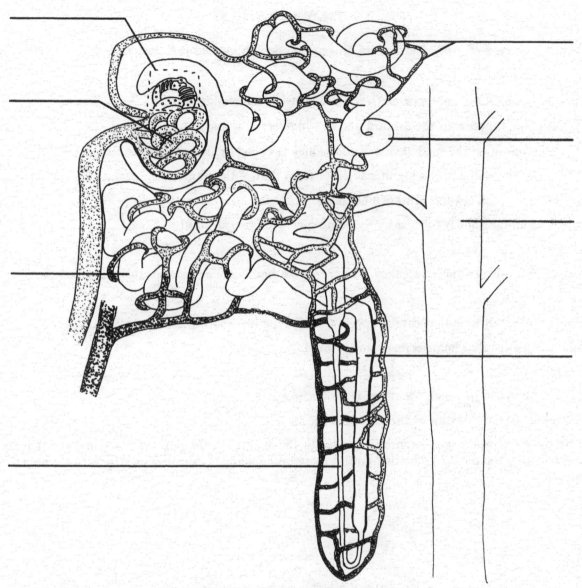

Figure 18-7. The nephron and its collecting duct.

Adjacent to the renal corpuscle are many cross sections of the proximal and distal convoluted tubules. Add one cross section of a tubule to your diagram. The wall of the tubule is composed of what type

of epithelial tissue? _____ The simple cuboidal epithelial cells play a very important role in reabsorbing substances from the lumen of the tubule back into the blood. If you look carefully at the field of vision you may see sections of the peritubular capillaries which wrap around the tubules. A section of capillary can be identified based on its single layer of endothelial (simple squamous) cells.

Move your mechanical stage until the field of vision displays the renal medulla. Note the presence of many tubules running lengthwise. These represent the limbs of Henle, collecting ducts and the vasa recta. Make a sketch of a small section of the medulla.

1. What is the sex of a pig that has its urogenital opening located just posterior to the umbilical cord? _____

2. The tube which transports urine from the kidney to the bladder is the _____ .

3. What is the function of the urethra and in which sex is it longer? _____

4. The depression on the medial side of the kidney is called the renal _____ .

5. How does the shape of the pig uterus differ from that of the human? _____

6. In the pig, which sex has a urogenital sinus? _____

7. The urogenital sinus is formed by the joining of what two structures? _____ and _____

8. When sperm leave the testes, they pass through a highly coiled mass of tubules called the _____ .

9. Name the small tube that carries sperm through the inguinal canal. _____

10. List the glands that help secrete seminal fluid. _____ , _____ , _____ .

11. Name the organ in which the ovum is produced. _____

12. The basic functional unit of the kidney is the _____ .

13. The metabolic waste product, urea is carried by the blood to the kidney. Trace the journey of a molecule of urea from the point of filtration until it reaches the bladder. Name each structure through which it passes.

VERTEBRATE PHYSIOLOGY: THE URINARY AND REPRODUCTIVE SYSTEMS

Objectives

1. Identify the location in the nephron where the process of filtration takes place and name at least four components found in blood that are filtered.

2. Perform tests on urine to determine the pH, specific gravity, and the presence of glucose, ketones, protein and phenylketo acids and interpret the results of each test.

3. Devise a procedure to determine the presence of the above named substances in an unknown sample of urine.

4. Identify substances that may be seen in a microscopic examination of urine sediment and explain the procedure required to prepare the sediment.

5. From a microscopic slide, identify the testis including interstitial cells and seminiferous tubules with spermatogonial cells, primary and/or secondary spermatocytes, spermatids and spermatozoa.

6. From a microscopic slide identify the ovary including primary follicle, mature or graafian follicle and corpus luteum.

7. Identify the pituitary hormones that control the ovary and testis and name the hormones produced by the growing follicle and corpus luteum.

8. Interpret the results of a pregnancy test and explain the procedure.

9. Answer all questions at the end of the exercise.

I. Introduction

In the previous exercise you had the opportunity to examine the anatomy of the urinary and reproductive systems. In this exercise you will investigate how these two systems function.

II. Kidney Physiology

The kidneys perform a variety of vital functions, including helping to maintain normal blood volume, regulating water and salt balance and removing many metabolic waste products. The kidney accomplishes these tasks by filtration, reabsorption and secretion.

Every minute the heart pumps over 5000 ml of blood throughout the body. Of this amount (the cardiac output), over 1000 ml flows through the kidneys every minute. As the blood flows into the glomerulus of each nephron, a portion of the blood plasma (125 ml/min) passes from the glomerulus into the Bowman's capsule by the process of filtration. Only those substances small enough to flow through tiny openings in the glomerulus and capsule wall will be able to enter the Bowman's capsule. Those substances that pass into the Bowman's capsule make up the *filtrate* which includes: water, ions, sugars, amino acids, and metabolic wastes such as urea, ammonia, creatinine and uric acid. Most plasma proteins are too large to be filtered. As the filtrate passes through the nephron tubules some components are reabsorbed from the tubule back into the blood, while other components continue through the tubules and collecting duct and enter the renal pelvis as urine. Urine also gains certain substances such as ammonia and hydrogen ions by secretion into the tubules.

Of the 125 ml of filtrate passing through glomeruli every minute, only 1 ml of urine is formed. Based on this figure, how many milliliters of urine would be formed in one hour?_____.
How many in one day? _____.

III. Urinalysis

An analysis of a urine sample reveals much information about kidney function and general body condition. Urinalysis usually includes tests for pH, specific gravity and for the presence of substances such as sugar, protein, ketones and blood. In a previous laboratory exercise (Chemical Aspects of Life) you had the opportunity to learn how to determine the pH of a substance and how to test for the presence of sugar and protein. In this exercise you will have the opportunity to learn some additional techniques, such as determining the specific gravity of a substance.

Specific gravity can be determined by using a hydrometer and hydrometer jar. Examine a hydrometer and locate the round ballast or "float" end. At the opposite end is the stem. Notice the specific gravity scale located within the hydrometer stem. The value at the top of the scale is 1.000. What is the value at the bottom of the scale? _____ What are the gradations of the scale, that is, what is the distance between each line? _____

Place 50 ml of tap water in a hydrometer jar and gently lower the float (hydrometer) into the water. Make sure the float does not touch the sides of the jar. Read the figure on the scale at the meniscus (the lowest portion of the water surface). Record the specific gravity of the water: _____.
The concept of specific gravity is used to compare the weight or mass of a substance with a standard. Pure water has a specific gravity of 1.000 and is used as a standard to which other liquids are compared. Is the water in the hydrometer jar pure water? _____ Liquids containing dissolved substances have a greater specific gravity than pure water.

What do you think is the specific gravity of 50 ml of water containing two grams of dissolved salt (NaCl)? _____. Using a laboratory balance, weigh two grams of NaCl and dissolve it in 50 ml of tap water. Stir the water until all of the NaCl is dissolved. Pour the solution into a hydrometer jar and determine its specific gravity. What is the specific gravity of the salt solution? _____. If additional NaCl were in solution, would the specific gravity increase or decrease? _____. Wash the hydrometer and jar.

In this portion of the exercise you will analyze unknown simulated urine specimens and a fresh actual urine specimen from yourself. Controls will be available for comparison.

After obtaining a fresh urine sample, describe its color:

_____ The normal color of urine ranges from "light straw" to amber. The color results from the presence of a pigment called urochrome. It is produced from the breakdown of hemoglobin by the liver, spleen and bone marrow during the normal destruction of old (120 days) red blood cells. Abnormal colors may result from foods in the diet, various diseases or from blood in the urine.

Next, record the appearance of the sample as clear or cloudy: _____ Fresh urine normally is clear, although by the end of the exercise it may become cloudy due to the precipitation of calcium phosphates. A cloudy fresh sample may be due to the presence of increased phosphates, carbonates, epithelial cells, mucus or a number of other factors.

Perform the tests indicated below and record them in Table 19-1.

A. *pH determination*

The pH of the freshly collected urine should be determined as accurately as possible using pH paper. Dip the pH paper into the urine to wet it. Touch the paper to the rim of the beaker to remove any excess urine and then compare the color of the pH paper to the color standards. If the pH meter is available a more accurate determination can be made. Record the results in the table 19–1. Urine normally has a slightly acidic pH of about 6, but may vary from 4–8 depending largely on diet. Why is the pH normally acidic? (Remember that the end products of cellular respiration include large amounts of CO_2 and lactic acid.)

B. Specific gravity determination

Fill the hydrometer cylinder about ¾ full with urine. Place the hydrometer in the cylinder and be sure it floats. If it touches the bottom add more urine. Spin the urinometer gently and then release it. When it stops it should *not* be touching the wall of the cylinder. Read the number that corresponds to the meniscus of the urine level. Record the specific gravity in table 19–1. Is

the specific gravity greater than or less than 1.000? _____

Why? _____

_____

The specific gravity of urine is normally between about 1.010 and 1.030, but may be higher with excessive sweating or lower with increased fluid intake. In the disease condition diabetes insipidus the urine is unusually dilute. This disease is usually caused by hyposecretion of antidiuretic hormone (ADH) which regulates the reabsorption of water by the collecting ducts. Record your results in the table 19–1.

C. *Glucose determination*

Glucose may be determined by several methods, including the Benedict's test. Obtain three test tubes and label them 1–3. In tube one place 3 ml of the known glucose solution; in tube two place 3 ml of the urine specimen; in tube three place 3 ml of tap water. To each add an equal amount of Benedict's reagent. Place the tubes in boiling water for three minutes and then remove. A color change indicates the presence of a reducing sugar such as glucose. Record the results of the glucose test in the table as present or absent.

For many years methods have been available for the rapid determination of glucose in urine for home use by diabetics. These consist of dipsticks with reagents on them that change color when wet and exposed to glucose. Such a rapid test for glucose will be performed with the test for ketones. Glucose is not normally found in detectable amounts in normal urine. Its presence may indicate a diabetes mellitus condition resulting from inadequate amounts of insulin production by the pancreas.

D. *Protein determination*

Place about 5 ml of the urine in a test tube and label it. In another tube place about 5 ml of the albumin (protein) solution. Immerse both in boiling water for at least five minutes. If a white precipitate forms add 5 drops of 2% acetic acid solution to the tube. If the precipitate persists then protein is present. If no precipitate forms, or if it disappears when the acetic acid is added, the sample contains no protein. Record your results in the Table 19–1.

Proteins are large molecules that are not normally filtered out of the blood in the glomeruli and thus are not normally found in the urine. In kidney disease proteins, and particularly the smallest blood proteins, the albumins, may be found.

TABLE 19-1

Test	Observation
pH	
Specific gravity	
Glucose	
Protein	
Ketones and Glucose	

E. *Ketone and glucose determination*

Perform this test using the Keto-diastix® reagent strips. Read the instructions on the bottle first. Remove two sticks from the container and close the container again immediately. Do not touch the reagents on the stick. Dip one stick into the known ketone solution. Wait exactly 15 seconds and then compare the stick to the color standard for ketone determination on the bottle. Repeat using the second stick for the urine sample. At exactly 30 seconds read the glucose test results also. Record your observations in Table 19-1. Ketones may appear in the urine of diabetics and of people on starvation or low carbohydrate diets.

F. Phenylketo acids

Place 8 drops of 5% ferric chloride ($FeCl_3$) in each of two test tubes. Label one tube "control" and add 5 drops of PKU urine, mixing gently. To the second tube add 5 drops of an unknown urine sample and mix gently. After 90 seconds, record the color in each tube.

Tube	Color
PKU CONTROL	
UNKNOWN: _____	

The ferric chloride reacts with keto acids such as phenylpyruvic acid producing a green or grey-green color. Phenylpyruvic acid is present in the urine of individuals with the inherited (autosomal recessive) metabolic disorder, phenylketonuria (PKU). The lack of a particular enzyme prevents the proper metabolism of the amino acid phenylalanine, resulting in a build up of phenylalanine in the blood. The ultimate consequence of the build up results in mental retardation. Evidence of high levels of phenylalanine can be detected by the presence of phenylpyruvic acid (a metabolite of phenylalanine) in the urine.

G. Analysis of unknown samples

Your instructor will assign to you one or more unknown urine samples to analyze. Perform the analysis, record your findings in the chart below and then interpret them.

216

Unknown Designation:	A	B	C	D
APPEARANCE: 1. color				
2. clarity				
TESTS: 1. pH				
2. specific gravity				
3. glucose determination				
4. protein determination				
5. ketone determination				
6. phenylketo acid determination				

INTERPRETATIONS OF RESULTS:

For each sample analyzed, formulate a possible explanation of any observation which is not consistant with a normal urine sample.

H. *Microscopic examination of urine*

Place 10 ml of the urine sample in a centrifuge tube and centrifuge for 5–10 minutes. Pour off most of the supernatent liquid and place some of the sediment on a slide. Add a drop of Sedistain® if it is available, coverslip and examine. Compare your findings with the figures in the Ames Atlas of Urine Sediment (see page 218). Numerous epithelial cells and crystals are normally found in urine. Numerous blood cells and bacteria are not normal and may indicate renal pathology.

Clean up all lab materials before you leave. The urine samples may be discarded in the sink. Glassware is to be washed in the soap solution provided and then rinsed and placed back on the lab table.

CRYSTALS FOUND IN ACID URINE 400 X

| Uric acid | Amorphous urates and uric acid crystals | Hippuric acid | Calcium oxalate | Tyrosine needles Leucine spheroids Cholesterin plates | Cystine |

CRYSTALS FOUND IN ALKALINE URINE 400 X

| Triple phosphate Ammonium and magnesium | Triple phosphate going in solution | Amorphous phosphate | Calcium phosphate | Calcium carbonate | Ammonium urate |

SULFA CRYSTALS

| Sulfanilamide | Sulfathiazole | Sulfadiazine | Sulfapyridine |

Ames Atlas of Urine Sediment

Printed by permission
Ames Division
Miles Laboratories, Inc.
Elkhart, Indiana 46515

CELLS FOUND IN URINE

| RBC and WBC | Renal epithelium | Caudate cells of Renal Pelvis | Urethral and bladder epithelium | Vaginal epithelium | Yeast and bacteria |

CASTS AND ARTIFACTS FOUND IN URINE 400 X

| Granular casts fine and coarse | Hyaline cast | Leukocyte cast | Epithelial cast | Waxy cast | Blood cast |

| Cylindroids | Mucous thread | Spermatozoa | Trichomonas vaginalis | Cloth fibers and bubbles |

IV. Reproductive Physiology

A. The Testis (plural = testes)

The testes produce sperm and the major male sex hormone, testosterone. The activity of the testes is regulated by two pituitary hormones:

1. FSH (follicle stimulating hormone)—stimulates sperm production

2. LH (luteinizing hormone)—stimulates the production of testosterone

Examine a slide of a cross section through the testis of a mammal. The structures within the testis that produce sperm are the *seminiferous tubules*. The tubules are very convoluted and therefore appear as circular or somewhat elongated. Center one section of a seminiferous tubule and switch to high power. The cells at the periphery of the tubule are *spermatogonial cells*. These cells undergo cell division (mitosis) to produce additional spermatogonial cells and cells called *primary spermatocytes*. The large primary spermatocytes undergo meiosis I to produce smaller *secondary spermatocytes*, which in turn undergo meiosis II to form tiny *spermatids*. The spermatids then mature into spermatozoa (sperm cells). Draw a pie-shaped section of the seminiferous tubule and label a spermatogonial cell (the cells at the periphery), a primary and/or secondary spermatocyte (the cells just interior to the spermatogonial cells) and a spermatid (very small cells near the lumen or central cavity of the tubule). Sperm cells can be identified by their very long thin tails which are directed toward the lumen.

In addition to sperm, what else does the testis produce? _____. The cells that produce testosterone are called *interstitial cells* and are located in the tissue spaces in between the tubules. Add two or three interstitial cells to your drawing.

B. Ovary

Examine a slide of a section through the ovary of a mammal. In each ovary numerous oocyte containing follicles can be seen in various stages of development. The pituitary hormones, FSH and LH control the development of the follicles.

Just inside the margin of the ovary are many primordial follicles, each consisting of a developing oocyte surrounded by single layer of flattened cells. Upon stimulation by FSH, several follicles will develop into *primary follicles*, in which the follicular cells become cuboidal. Development continues as primary follicles become *secondary follicles*. Secondary follicles are characterized by having several layers of cells surrounding the oocyte and eventually developing fluid filled spaces. Continued stimulation of FSH results in a mature or *graafian follicle*. The graafian follicle is very large. The oocyte, surrounded by follicular cells, usually is located to one side. Because the graafian follicles are so large, the oocyte may not be present in some of the sections. In the space below draw a primary follicle, secondary follicle and a graafian follicle. Label the oocyte and the follicular cells.

Developing follicles produce the estrogen hormones which trigger the release of LH. The combined influence of FSH and LH results in the eventual rupturing of the graafian follicle and the release of the oocyte from the ovary (ovulation). The ruptured follicle collapses and is transformed into a large *corpus luteum* ("yellow body"). The corpus luteum secretes estrogen and another hormone called progesterone. Both of these hormones help to prepare the body for a possible pregnancy. Examine the slide labeled "corpus luteum" under low power and draw the corpus luteum.

C. Pregnancy Test

If fertilization occurs with subsequent uterine implantation, certain embryonic cells (the trophoblast) begin to produce a hormone called *human chorionic gonadotropin* (HCG). HCG passes into the maternal circulation and helps maintain the corpus luteum. This ensures that ample supplies of estrogens and progesterone will be available until the developing placenta is capable of its own hormonal production. Excess HCG passes out of the mother through the urine and provides the basis of the pregnancy test.

If materials are available, your instructor will demonstrate the testing procedure.

Negative Control

1. Place one drop of anti-HCG serum on a special glass slide. The anti-serum contains antibodies that bind to HCG.
2. Add urine that is known NOT to contain HCG (negative control) and mix it with the anti-serum.
3. Add one drop of an HCG-coated latex suspension (latex particles that are coated with HCG). What interaction is expected to occur between the anti-HCG antibodies and the-HCG coated latex particles?

_____

What visible evidence would be expected to be seen when the antibodies bind to the latex particles?

_____

4. Agglutination (clumping) of the latex particles is indicative of a negative test for pregnancy.

Positive Control

Repeat steps 1 through 3 except in step 2, the urine sample contains HCG.

What would be expected to occur when the urine containing HCG is mixed with the anti-HCG antibodies? _____

_____

Since HCG molecules are very small, would the binding of the antibodies to the HCG molecules be visible? _____

When the HCG-coated latex particles are added, would you expect any of the antibodies to be available to react with the HCG coated latex particles? _____ Would

agglutination of the latex particles be expected to occur? _____

The absence of any agglutination of latex particles indicates a positive test for pregnancy.

REVIEW QUESTIONS

1. In the nephron, substances are filtered from the _____ into _____'s

   _____ .

2. Which of the following are too large to be filtered? urea, simple sugars, proteins, red blood cells, white blood cells, amino acids, electrolytes (salts, etc.).

   _____

3. Which of the following pH values BEST represents a pH within the normal range for a fresh urine specimen? 2, 6, 9. _____

4. When a sample of urine is not fresh and has not been refrigerated, the urea breaks down into ammonia and carbon dioxide, with most of the carbon dioxide escaping into the air. How will this affect the pH value?

   _____

5. In a starvation diet increased metabolism (breakdown) of fats and protein occurs. In what way would this affect the pH of a urine sample? Explain. _____

6. With excessive sweating, will the specific gravity of urine increase or decrease? Why? _____

   _____

7. In the disease condition diabetes mellitus, what substances appear in the urine which do not appear in normal urine?

   (a) _____ (b) _____

8. Why are proteins not normally present in the urine?

   _____

9. Identify the function of the interstitial cells.

   _____

10. Which of the following cells is expected to be the smallest: spermatid, oogonial cell, primary spermatocyte?

11. In what ways does a graafian follicle differ from a primary follicle? _____

12. Name the two pituitary hormones which influence follicular development. _____

13. A sample of urine is tested for the presence of HCG utilizing the technique mentioned in the exercise. At the end of the test agglutination is seen. Interpret the result of this test.

ANIMAL DEVELOPMENT

Objectives

Upon completion of this exercise you should be able to:

1. Distinguish between direct and indirect development.
2. Identify all of the stages in a complete metamorphosis pattern of development.
3. Trace the development of a starfish from egg to gastrula stage.
4. Trace the development of a frog from egg to neural tube stage.
5. Contrast the eggs and developmental patterns among starfish, frog, and chick.
6. Trace human embryonic development using models and identify morula, neural tube, heart, gill (branchial) arches and pouches, somites, and limb buds.
7. Identify and give the functions of the chorion, amnion, allantois, yolk sac, and placenta.
8. Identify the major developmental events that occur in each trimester period of human development.
9. Answer all of the review questions.

I. Introduction

The development of a sexually reproducing organism begins with one cell (the fertilized egg or zygote) and eventually ends as an adult organism. The entire process involves cell division, cell movement, cell growth, and cell specialization. Animals which undergo direct development (reptiles, birds, mammals) pass from the embryonic state into a form which basically resembles the adult, except on a much smaller scale. Animals which undergo indirect development proceed through one or more intermediate forms before reaching the sexually mature adult. In some cases the change in body form (metamorphosis) is complete (i.e. transformation of caterpillar into moth), while in others, the metamorphosis is incomplete (transformation of wingless grasshopper into winged grasshopper).

In this laboratory exercise you will observe the complete metamorphosis pattern in a fruit fly and examine the initial steps of development in a starfish, frog, and chick. You also will trace the developmental pattern in the human.

II. Complete Metamorphosis in the Fruit Fly

The complete metamorphosis pattern of the fruit fly begins with an egg which hatches into an immature feeding stage called a larva. The larva eventually develops into a nonfeeding stage called a pupa. Within the pupa stage the larval tissue is broken down and adult tissue begins to form. The adult tissue forms from isolated groups of cells (imaginal discs) which have been set aside during early embryonic development. Finally the adult organism (imago) emerges from the pupa casing. Observe and diagram each of the stages in the development of the fruit fly, *Drosophila melanogaster*. Use the stereoscopic microscope to aid in locating and identifying the various phases.

EGG STAGE: The eggs are tiny (0.5mm) white oval-shaped structures with two flotation devices at the anterior end. Look for the eggs on the surface of the culture medium. Draw an egg in the space below:

LARVAL STAGES: The fruit fly goes through three larval stages, with each referred to as an instar. The first instar (1.5mm) lasts for about 24–36 hours, then sheds its outer covering (molts) and grows into a very active second instar (3mm). Within 48–72 hours a second molt occurs resulting in a slow moving third instar (5mm).

Place a second or third instar on the stage of a stereoscopic microscope and, by placing a light source beneath the instar, identify the well developed mouth hooks and the tracheal tubes. The two tracheal tubes run the length of the body and open to the exterior by means of spiracles. What is the function of the tracheal tube system? _____ _____ Draw one instar and label mouth hooks and tracheal tubes.

PUPAL STAGE: The third instar travels to a relatively dry region and develops into the pupa. The covering (cuticle) of the instar darkens and hardens to form the pupal casing. What activity occurs within the pupa? _____ Where does the nourishment (energy) for this activity come from? _____ Draw a darkened pupa and identify:

   a. two anterior "pupal horns" which are evaginations of the anterior spiracles. What is their function? _____
   b. two large red eyes developing within the pupal casing.
   c. two folded wings just posterior to the eyes on the dorso-lateral side of the casing.

The eyes, wings, and other adult tissues develop from isolated groups of cells called _____
_____ .

III. **The Early Development of the Starfish (Fig. 20–1)**

The starfish has been selected to illustrate the early phases of the developmental process because:
   1. The general pattern of development is similar to that in the human, but is less complicated.
   2. The egg and embryo are transparent.
   3. The egg contains very little yolk material.

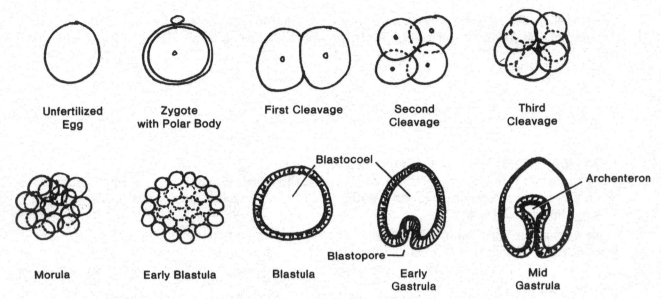

Unfertilized Egg	Zygote with Polar Body	First Cleavage	Second Cleavage	Third Cleavage
Morula	Early Blastula	Blastula	Early Gastrula	Mid Gastrula

Figure 20-1. Starfish development.

View the prepared slide of *Starfish Development,* and identify each phase listed below. Make a sketch of each phase.

1. UNFERTILIZED EGG—The very small amount of yolk in the egg is distributed evenly throughout the cytoplasm. This type of egg is called *isolecithal.* What can you infer about the length of the development period based on the amount of stored yolk? _____

2. ZYGOTE—After fertilization, the membrane around the egg undergoes a change and forms a fertilization membrane. This prevents additional sperm from fertilizing the egg. The zygote (fertilized egg) may be differentiated from the unfertilized egg by the presence of the fertilization membrane and, in some cases, by the presence of two polar bodies located just outside of the zygote.

3. FIRST CLEAVAGE—The first cleavage furrow begins to divide the zygote into two small cells called *blastomeres.* Is the first cleavage vertical or horizontal?

4. SECOND CLEAVAGE—The second cleavage also is vertical, but is at right angles to the first. When viewed from above, four blastomeres may be seen.

224

5. THIRD CLEAVAGE—The third cleavage is horizontal along the equator of each blasto-mere, resulting in four blastomeres on top and four beneath. Compare the size of the top and bottom cells.

6. MORULA—When the embryo reaches 16 cells it is a solid ball of cells called a morula.

7. BLASTULA—As blastomeres continue to divide, a central hollow cavity *(blastocoel)* begins to form. The resulting hollow ball of cells is called a *blastula*. Compare the size of the blastula to that of the original egg.

8. EARLY GASTRULA—The *gastrula* begins to form when cells at one end of the blastula begin to migrate inward. Find an early gastrula and note the infolding. The opening pro-duced by the infolding is called the *blastopore*.

9. MID GASTRULA—As cells continue to migrate inward, what happens to the size of the original blastocoel? _____ As the blastocoel is reduced in size, a new cavity, the *archenteron,* begins to form. The archenteron opens to the exterior through the blastopore. The archenteron will become the primitive gut of the embryo and the blastopore will become the anal opening. Only in echinoderms (starfish, sea urchins, etc.) and chordates (vertebrates, etc.) does the blastopore become the anus. In all other animal phyla the blastopore becomes the mouth.

10. MID TO LATE GASTRULA—Find a gastrula on the slide in which the archenteron is well formed. The outermost layer (*ectoderm*) covers the gastrula and gives rise to the body covering and nervous system. The layer of cells lining the interior of the archenteron is the *endoderm* and gives rise to the lining of the digestive tract. In between the two layers, a middle layer *(mesoderm)* will form. The cells that will form the mesoderm may be seen scattered throughout the old blastocoel. The mesoderm gives rise to muscle, skeletal, and reproductive structures.

11. BIPINNARIA LARVA—The starfish gastrula develops into a ciliated free swimming larval form (*bipinnaria*) which later metamorphoses into the radially symmetrical adult.

If models are available, identify all phases and structures in starfish development from zygote to gastrula.

IV. Development in the Frog (Fig. 20-2)

A. The Frog Egg

View the video "Amphibian Embryo" or use your textbook as a reference for the following section.

1. Contrast the frog's *mesolecithal* egg with the starfish's isolecithal egg in reference to amount of yolk and yolk location.

2. Which portion of the frog's egg possesses the darker pigmentation?

3. What is the upper darker portion of the egg called? _____ pole

4. The vegetal pole is much heavier than the pigmented animal pole due to the presence of _____ .

5. Frog eggs are fertilized externally and are left alone to develop in the water. What survival value is obtained from the distribution of pigment and yolk?

6. After fertilization, when the upper pigmented layer rotates toward the point of sperm entry, the region of cytoplasm on the opposite side becomes known as the _____ _____ .

B. Cleavage Through Gastrulation

Examine the slides: *Frog Late Cleavage sec., Frog Blastula sec.,* and *Frog Yolk Plug sec.*

1. Contrast the size of the blastomeres at the vegetal pole with those at the animal pole. How can you account for the difference in size?

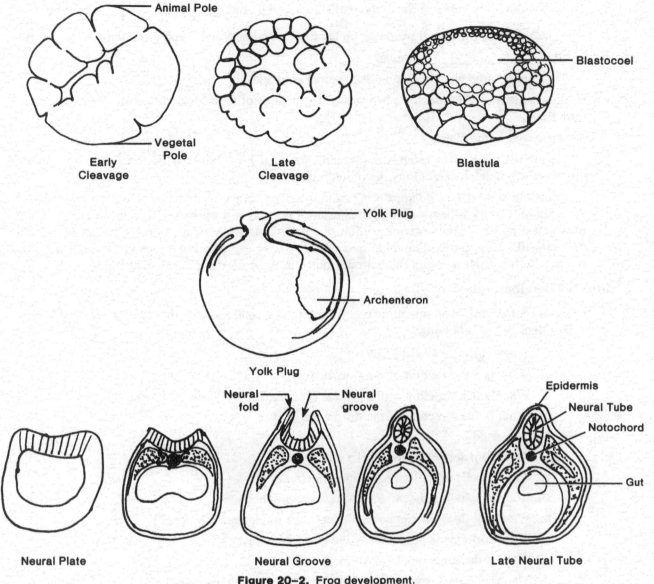

Figure 20–2. Frog development.

2. The cavity within the frog blastula is called the _____ .

3. The slit-like opening around the yolk plug is called the dorsal lip of the _____ .

4. The opening through the dorsal lip of the blastopore leads to a newly formed cavity called

    the _____ .

5. Identify the three embryonic germ layers formed in the gastrula.

C. Neural Plate Through Neural Tube

Examine the slides: *Frog Early Neural Groove c.s.*, *Frog Late Neural Groove c.s.*, and *Frog Late Neural Tube c.s.* (Note: These slides are cross sections and the previous slides were saggital sections). Trace the development of the neural tube through each section.

1. The thickened layer of dorsal ectoderm is known as the _____ plate.

2. The neural plate is induced by an underlying mesodermal supportive structure called the

    _____ .

227

3. In almost all vertebrates the notochord is replaced by the _____ column.

4. The ridges of the neural plate rise up (to form a neural groove) and eventually meet to form the _____ tube.

5. The neural tube will develop into the spinal cord and _____.

If models are available identify all stages and structures of frog development from egg to neural tube stage.

D. Development in the Chick

Use your textbook as a reference (organogenesis in bird embryo) or the following URI http://www.uoguelph.ca/zoology/devobio/dbindex.htm.

Examine the slide. *Chick 48 hour w.m.* using the stereoscopic microscope. Measure the length of the embryo. Contrast the position of the posterior portion of the embryo with the anterior portion. In which direction is the anterior portion oriented? Note the size and bending of the brain. The dark cup-like structure will develop into what? _____. The large organ bulging toward the right is the heart. Now locate the paired somites in the posterior portion of the embryo.

V. Human Development

View a film on human development or use your text as a reference for this section. If models are available, identify the following:

1. Cleavage at the 2, 4, and 8 cell stages—
   a. What is the direction of the first cleavage?
   b. What is the direction of the second cleavage?
   c. What is the direction of the third cleavage?

2. Morula stage—
   a. Approximately how many cells are present?
   b. Are the cells of the same or different sizes?

3. Blastocyst including the inner cell mass and trophoblast—
   a. What role does the trophoblast play in the implantation process?
   b. What hormone is produced by the trophoblast layer?
   c. What is the function of the hormone?
   d. The trophoblast develops into which extraembryonic membrane?

4. Two week stage including neural folds, neural groove, closed neural tube and paired somites—
   a. Which end is the anterior end? Is there any evidence of an enlargement at the anterior end of the neural groove? This enlarged region will develop into what organ?
   b. Into what types of tissue will the somites develop?

5. Four week stage including heart and gill (branchial) arches and pouches—
   a. Is there any evidence of heart formation at the two week stage?
   b. At what day does the heart begin to beat?
   c. The first branchial arch gives rise to the lower jaw and the remaining arches contribute to the formation of the hyoid bone, laryngeal cartilage, and the bones of the middle ear. One of the gill pouches continues as the eustachian tube, connecting the pharynx with the middle ear.

6. Five week stage including limb buds, tail and brain—

    a. Which pair of limb buds appears to develop most rapidly?

    b. What evolutionary explanation can you give for the presence of a tail?

    c. The brain consists of how many major regions at this stage of development?

7. Six week stage with embryonic membranes shown—

    a. Name and locate the membrane which forms the embryonic portion of the placenta.

    b. Name and locate the membrane that forms a sac around the embryo.

    c. What is the function of this fluid filled sac?

    d. What, if any, is the function of the yolk sac?

8. Eight week stage with uterus shown—

    a. Locate the placenta. What comprises it?

    b. What is the function of the placenta?

9. After eight weeks head, trunk and limbs are evident and all organ systems have begun development—

    a. What is this stage of development called?

    b. What is the role of the umbilical cord?

Make a list of the major developmental events that occur in each trimester period:

REVIEW QUESTIONS

1. Distinguish between direct and indirect development.

2. Make a drawing of a fruit fly egg, larva and pupa.

3. How many fruit fly instar stages are there?

4. What is the major activity of the fruit fly larval stage?

5. What major activity occurs during the pupa stage?

6. Draw a starfish blastula and gastrula and label blastocoel, blastopore, and archenteron.

7. Name the three fundamental embryonic tissue layers and indicate the types of tissues or structures they produce.

8. Describe the difference among the starfish egg, frog egg, and bird egg.

9. The neural tube develops into the _____ and the _____ _____ .

10. What is the function of the trophoblast?

11. Name and give the function of the four extraembryonic membranes in human development.

12. Match each statement about development with the correct term.

_____ Site at which nutrients, gases and wastes are exchanged between the mother and developing embryo/fetus.

_____ Transports blood between the placenta and developing embryo/fetus.

_____ Stage that implants in the lining of the uterus.

_____ Stage during which all organ-systems start to develop.

A. blastocyst

B. embryo

C. fetus

D. placenta

E. umbilical cord

NERVOUS CONTROL:
NERVE CENTERS AND EFFECTORS

Objectives

Upon completion of this exercise you should be able to:

1. Label the parts of a motor neuron (figure 21–1) and give the function of the myelin sheath.
2. Identify and give the function of each of the five parts of a typical reflex arc.
3. List, give the functions of and be able to distinguish between the divisions of the nervous system.
4. Identify the term meninges.
5. Distinguish between white and gray matter.
6. Identify from a diagram, model or specimen the following regions of the brain and give the functions of each: medulla, pons, cerebellum, hypothalamus, thalamus, cerebrum.
7. Distinguish between cranial and spinal nerves.
8. List three functions of the skeletal system.
9. List the two major divisions of the skeletal system and the main parts of each.
10. List and identify by example the major types of articulations which occur in the human skeleton.
11. List the major groups of vertebrae and explain how each group is specialized for its particular functions.
12. Identify the atlas and axis and give the functions of each.
13. Distinguish among muscle origin, insertion and action.
14. If given the origin and insertion of a muscle, give its specific action.
15. List four major types of muscle action and give an example of each.

In the first part of this exercise, you will examine the basic nerve cell (neuron), transmission of the impulses and their destinations in the nervous system. In the second part you will examine the skeletal and muscle systems as examples of the action of effectors.

I. The Neuron: Structure and Function

A. Neuron Structure

The basic message conducting cell of the nervous system is the *neuron*. All neurons fundamentally are alike, consisting of a *cell body,* one or more processes called *dendrites,* and a single process called an *axon.* Dendrites carry impulses toward the cell body, while the single axon carries impulses away from the cell body. In Figure 21–1 label the cell body, nucleus, dendrites, and axon.

The processes of some neurons possess a covering called a *myelin sheath.* This sheath serves to insulate the neuron and allows for faster conduction of the nerve impulse. The myelin sheath of nerve cell processes located outside of the brain and spinal cord is produced by cells called *Schwann cells.* The Schwann cells play a vital role in the regeneration of damaged nerve cell processes. Neurons located within the brain and spinal cord may possess a myelin sheath, but the sheath is not produced by Schwann cells. Because of this difference damaged processes within

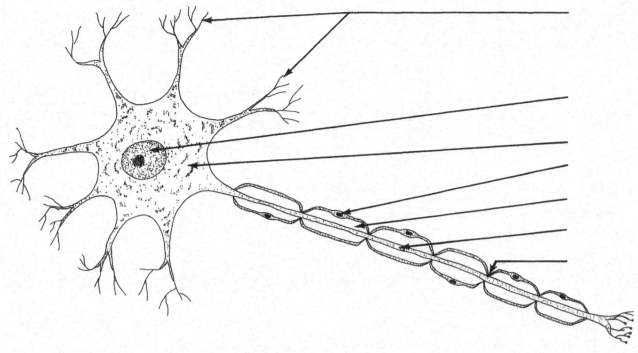

Figure 21–1. Motor neuron.

the brain and spinal cord are not capable of regeneration if they are cut or damaged. Label the myelin sheath and a Schwann cell in Figure 21-1. Note that where one Schwann cell meets an adjacent Schwann cell the myelin sheath is absent. These points are called *nodes of Ranvier.* Label a node of Ranvier in the figure, and identify all of the structures indicated in the figure on a model of the neuron.

B. *Conduction of a signal*—When the end of a nerve fiber is sufficiently stimulated, the stimulus initiates the chemical and electrical changes that travel over the length of the fiber. These changes are called the *nerve impulse.* The junction at which the axon of one neuron meets a dendrite of another neuron is called a *synapse.* At the synapse the "nervous signal" must be chemically transmitted across the *synaptic cleft.*

C. *Reflex Behavior and Reflex Arc*—

1. A *reflex* is an automatic response to a stimulus. The specific path the nerve impulse takes is a *reflex arc.* There must be at least two neurons in a reflex arc, but usually there are more. A typical reflex arc usually consists of:

    a. A *receptor* which is sensitive to environmental stimuli.

    b. A *sensory neuron* which carries the impulse toward a centralized nerve center (i.e., brain, spinal cord).

    c. A *nerve center* where synaptic junctions are made between the sensory neuron and interneurons.

    d. A *motor neuron* which makes synaptic junction with interneurons and carries the impulse toward the effector.

    e. *An effector* (i.e., muscles and glands) which brings about a response.

2. *Demonstration of a simple reflex.*

    a. *The patella reflex*—the patella reflex is best demonstrated by using two people although it is possible to perform it yourself. Have your partner sit down and cross his/her legs. Have your partner stare at the ceiling and count backwards from one hundred. While he/she is doing this, sharply tap the tendon just below the knee cap. This should result in an extension of the lower leg. Next have your partner stare at his/her leg and concentrate on it. Try to obtain the patella reflex under these conditions. Is there any difference?

(1) In the first case, the sharp tap on the tendon causes it to stretch. The stretching serves as a stimulus to the sensory nerve endings located in the tendon.

The impulse travels along the _____
(dendrite, axon) of the _____ (motor, sensory) neuron to the cell body. In this case, the cell body of the neuron is located in a ganglion (collection of nerve bodies) just outside of the spinal cord. The impulse then travels

from the cell body over the sensory (dendrite, axon) _____

into the spinal cord. Within the spinal cord the impulse jumps across a _____

junction onto the dendrite of a _____
(sensory, motor) neuron. The impulse then travels to the cell body of the motor neuron

and out across its _____ (dendrite, axon). The

axon leaves the _____cord and continues into the extensor muscle above the knee. The impulse causes the extensor muscle to contract, thus extending the lower leg.

During this time, impulses are jumping synapses with interneurons. Impulses travel along interneurons to the brain allowing the individual to be aware that his/her tendon was tapped. Find the above named structures on Figure 21-2.

(2) In the second case the individual sent impulses down *from* the brain to flexor muscles which have the opposite action as the extensor. In this manner, the individual is able to reduce the patella reflex or even suppress it altogether.

II. Central Nervous System

The central nervous system is composed of the brain and spinal cord. In all vertebrates the brain is relatively well developed and is continuous with the dorsal, hollow, fluid-filled single spinal cord. Both the brain and spinal cord are surrounded by three layers of connective tissue collectively referred to as *meninges*.

Spaces between the protective layers (meninges) are filled with cerebrospinal fluid forming a liquid protective cushion around the brain and spinal cord.

A. *Spinal Cord*—Examine a cross section of the spinal cord of a mammal. Identify the inner H-shaped zone of *gray matter* composed largely of the cell bodies of neurons, and the outer zone of *white matter* consisting of the myelinated processes of neurons. What is the function of the

myelin sheath? _____

A *central canal* filled with cerebrospinal fluid is located in the center of the gray matter. This canal results from the unique development of the chordate nervous system which includes a dorsal hollow nerve cord. Refer to Figure 21-2 and label gray matter, white matter and central canal.

B. *The Brain*—The brain also consists of both white and gray matter, except the gray matter is on the outside. Observe the relationship between the white and gray matter in a section of a human brain embedded in plastic.

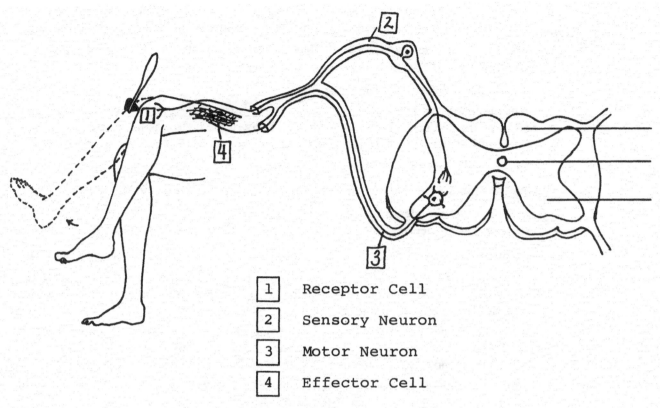

1	Receptor Cell
2	Sensory Neuron
3	Motor Neuron
4	Effector Cell

Figure 21-2. Simple reflex.

Examine a model of the human brain and a preserved sheep brain. Locate the following major areas using Figure 21-3 as a guide.

1. *Medulla*—Locate the brain stem area where the spinal cord begins. Just above this is the area of the medulla. The medulla consists of a large amount of white matter throughout which is scattered small amounts of gray matter.

   a. The white matter consists of many nerve cell fibers which *conduct impulses* from the spinal cord to the brain and from the brain back down the spinal cord eventually to effectors. Within the medulla many of the fibers crossover to conduct impulses to or from the opposite side of the brain.

   b. The gray matter consists of cell bodies which serve as *vital control centers*. These centers control the heart rate, the respiratory rate and help regulate blood pressure.

   c. The medulla also acts as a *reflex center* for swallowing, coughing, vomiting and sneezing.

2. *Pons*—Just above the medulla is a region consisting mainly of fibers running transversely (across) the brain stem. This is the pons area and it functions primarily as a *conduction* pathway for nerve fibers.

3. *Cerebellum*—Dorsal to, and overlapping the brain stem is the cerebellum. It contains both gray and white matter. The main function of the cerebellum is the *coordination* of muscular activity including the regulation of muscle tone, equilibrium and body posture.

4. *Hypothalamus*—Observe a saggital section of the brain. Locate the cavity (the third ventricle), just above the brain stem. The floor of this ventricle and part of the lateral walls comprise the hypothalamus. The hypothalamus functions as a *regulator of visceral (internal) activities* such as body temperature, water balance, appetite, and endocrine gland secretion.

Projecting down on a stalk from the hypothalamus is the pituitary gland. "Signals" from the hypothalamus to the pituitary stimulate the latter to release a variety of hormones.

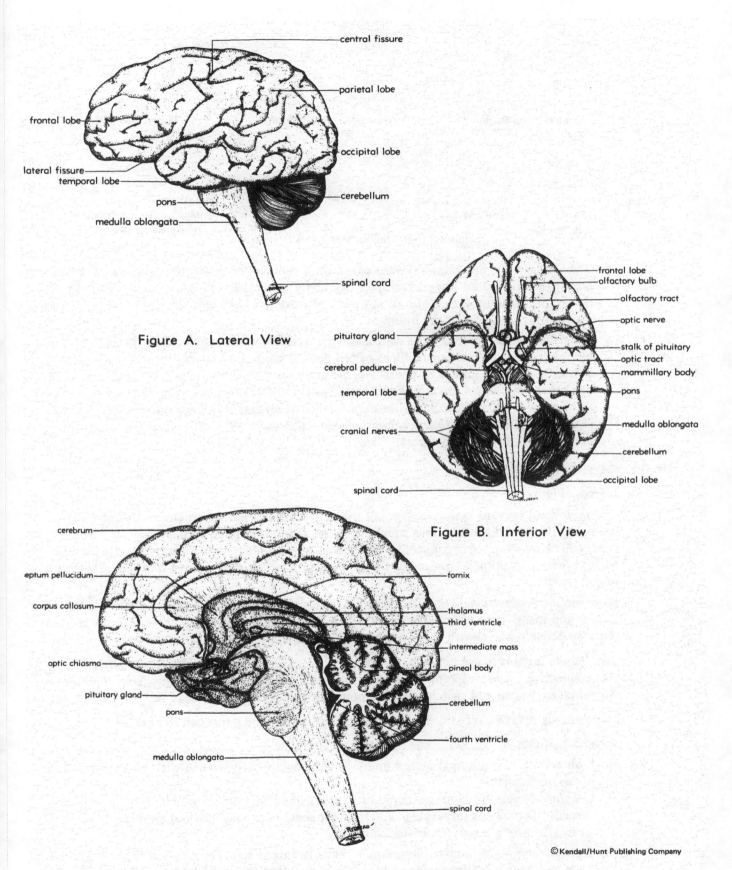

central fissure

parietal lobe

frontal lobe

occipital lobe

lateral fissure
temporal lobe

cerebellum

pons

medulla oblongata

spinal cord

Figure A. Lateral View

frontal lobe
olfactory bulb

olfactory tract

optic nerve

pituitary gland

stalk of pituitary
optic tract

cerebral peduncle

mammillary body

temporal lobe

pons

cranial nerves

medulla oblongata

cerebellum

spinal cord

occipital lobe

Figure B. Inferior View

cerebrum

fornix

septum pellucidum

corpus callosum

thalamus
third ventricle

intermediate mass

optic chiasma

pineal body

pituitary gland

cerebellum

pons

fourth ventricle

medulla oblongata

spinal cord

Figure C. Sagittal View

Figure 21-3. Human brain.

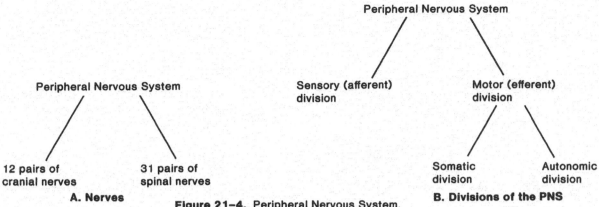

Figure 21-4. Peripheral Nervous System.

A. Nerves

B. Divisions of the PNS

5. *Thalamus*—The upper lateral walls of the third ventricle comprise the thalamus. The thalamus functions as an impulse *relay center* and a generalized *pain/pleasure center*. In addition, the thalamus *sorts out* all incoming sensory impulses (except smell) and redirects them to various parts of the brain.

6. *Cerebrum*—The cerebrum is a convoluted mass which occupies most of the brain area. The outer area (cortex) contains billions of cell bodies. Internal to the cortex is a mass of white matter containing groups of fibers which convey impulses to various regions of the *cortex* and other brain areas. *Discriminate sensations* (localized pain, sight, sound, etc.) reach *consciousness* in the cortex. Sensations are further *integrated* and *associated* with other sensations and memories. *Abstract thinking, speaking, voluntary actions* and *learning* all take place within the cerebral cortex.

III. Peripheral Nervous System

A. Nerves (Figure 21-4A.)

The peripheral nervous system consists of the paired cranial nerves which run to and from the brain and the paired spinal nerves which run to and from the spinal cord. A *nerve* is a group of nerve cell fibers wrapped in connective tissue. A nerve containing only sensory fibers is a sensory nerve; a nerve containing only motor fibers is a motor nerve; and a nerve containing both is a mixed nerve.

Examine the prepared slide of a peripheral nerve c.s and l.s. Using low power observe the cross section and identify the connective tissue which wraps around the entire nerve. Located within this connective tissue sheath are several bundles (fasciculi) of nerve cell processes. How many

bundles are present in the nerve on your slide? _____ Each bundle is surrounded by its own connective tissue wrapping. Within each bundle are many nerve cell processes. Center an individual nerve cell and switch to high power. The dark center consists of the nerve cell membrane and cytoplasm. What does the white sheath around the nerve cell represent? _____

What is its function? _____

Next, observe the longitudinal section and locate the nuclei of Schwann cells, the myelin sheath and nodes of Ranvier.

1. *Cranial Nerves*—in humans there are 12 pairs of cranial nerves which are primarily concerned with carrying messages to and from the sense organs, glands and muscles of the head. Cranial nerves are sensory, motor or mixed.

2. *Spinal Nerves*—in humans there are 31 pairs of mixed spinal nerves. Each nerve has two roots by which it is connected to the spinal cord. All of the sensory fibers enter the cord by the dorsal root, and all the motor fibers leave the cord by the ventral root. The nerve cell bodies of motor neurons are located in the ventral horns of the gray matter of the spinal

236

cord. The sensory nerve cell bodies are in the *dorsal root ganglia*, just outside the cord. Examine the model of a spinal cord with the spinal nerves.

B. Divisions (Figure 21–4B.)

The peripheral nervous system can be subdivided into the *sensory* (afferent) *division* which conducts sensory information to the central nervous system and the *motor* (efferent) *division* which carries impulses to effectors. Motor impulses in the *somatic division* lead to excitation of skeletal muscles. Motor impulses in the *autonomic division* lead to excitation and/or inhibition of cardiac muscle, smooth muscle, and various glands. Through its action the autonomic nervous system helps regulate the heart rate, blood pressure, digestive tract movement, etc.

IV. Effectors

An effector is anything that brings about a response. It can be a gland which secretes or a muscle which contracts. In this portion of the exercise you will study the action of skeletal (voluntary) muscles as an example of an effector. In higher animals voluntary muscles work in conjunction with the skeletal system. Therefore, you will examine the skeletal system first and then correlate it with the actions of the voluntary muscles.

A. Skeletal System

The skeletal system (Fig. 21–5, Fig. 21–6), is one of the most characteristic features of vertebrates. It is composed of bone and cartilage held together by ligaments, and functions for *support, muscle attachment* and *protection* of internal organs. Two major divisions of the skeletal system are generally recognized: The *axial skeleton*, consisting of the skull, vertebral column (backbone), ribs and sternum (breast bone) and the *appendicular skeleton*, which includes the fore- and hind-limbs and the girdles of bone which support them.

1. The Axial Skeleton.

   a. Skull—Using Figure 21–6 as a guide, examine the skulls on demonstration. Pay particular attention to the diverse ways in which bones connect with one another (joints or articulations). Almost all of the bones of the skull are joined together by very close articulations with neighboring bones. This type of joint is referred to as a *suture* and renders these bones immovable. In life a layer of fibrous connective tissue separates the two bones. Which one of the bones of the skull is not attached to its neighbors by sutures?

      _____.

      Place your finger just in front of the opening of your ear and open and close your mouth. You should be able to feel the *condyloid* process of your *mandible* (lower jaw) slide back and forth as it articulates with the *temporal* bone (the bone forming the lower portion of the side of the skull). Identify these bones on a demonstration skull and note the movement. The bony projections of the mandible serve as points of muscle attachment.

   b. Vertebral Column—Examine the vertebral column of the human skeleton. The first seven bones are the *cervical vertebrae*. Their function in addition to protecting the spinal cord is to permit movements of the head and neck.

      (1) Notice the first cervical vertebra, the *atlas*. It has been modified to articulate with the base of the skull and to allow the skull to tilt back and forward. The second cervical vertebra, the *axis*, has also been modified to allow the skull to turn to the right and left. The articulation between the atlas and axis is one of the *pivot joints* in the body.

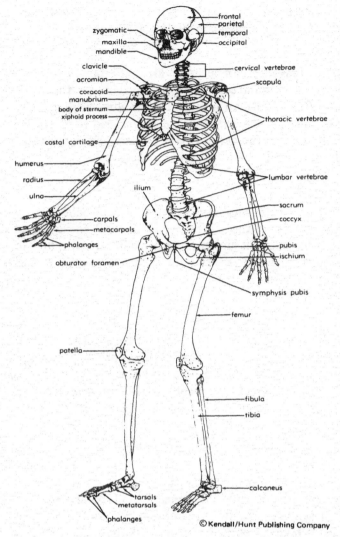

frontal
parietal
temporal
occipital
zygomatic
maxilla
mandible
clavicle
cervical vertebrae
acromion
scapula
coracoid
manubrium
body of sternum
xiphoid process
thoracic vertebrae
costal cartilage
humerus
radius
lumbar vertebrae
ulna
ilium
sacrum
coccyx
carpals
metacarpals
phalanges
pubis
ischium
obturator foramen
symphysis pubis
femur
patella
fibula
tibia
calcaneus
tarsals
metatarsals
phalanges

© Kendall/Hunt Publishing Company

Figure 21–5. Human, skeletal system, ventral view.

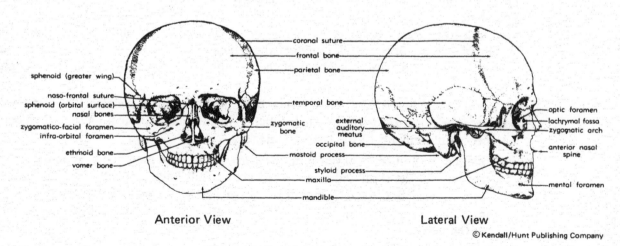

coronal suture
frontal bone
parietal bone
sphenoid (greater wing)
naso-frontal suture
sphenoid (orbital surface)
nasal bones
temporal bone
zygomatico-facial foramen
infra-orbital foramen
zygomatic bone
external auditory meatus
optic foramen
lachrymal fossa
zygomatic arch
ethmoid bone
occipital bone
vomer bone
mastoid process
anterior nasal spine
styloid process
maxilla
mandible
mental foramen

Anterior View
Lateral View

© Kendall/Hunt Publishing Company

Figure 21–6. Human, skeletal system, skull series.

(2) Place your fingers on the back of the neck at the base of the skull. Move them slowly down your neck until you reach a large bump (nape of neck). Identify this "bump" on the skeleton. This represents the elongated *spinous process* of the seventh cervical vertebra. The various processes on all of the vertebrae function as muscle attachment points.

(3) The next twelve vertebrae, the *thoracic*, are modified to allow articulation with the ribs.

(4) The next five vertebrae, the *lumbar*, bear much of the weight of the column and are modified for attachments of the lower back muscles.

(5) The next five vertebrae are fused to form the *sacrum* which articulates with the hip bones.

(6) The last four rudimentary vertebrae are fused to form the *coccyx*.

(7) Note that *all* vertebrae possess a large opening, the *vertebral foramen*, through which the spinal cord passes. The various processes of the vertebra are used in articulating with other vertebrae or as muscle attachment points. All the vertebrae except the atlas and axis possess a thick rounded *body*. Between the bodies of adjacent vertebrae are *intervertebral disks* of fibrous cartilage. These disks act as a cushion and shock absorber. The articulations between these vertebrae are classified as slightly moveable joints (*amphiarthroses*).

c. List two other components of the axial skeleton besides vertebrae and locate them on the skeleton.

(1)

(2)

2. Appendicular Skeleton—The appendicular skeleton consists of the appendages and girdles. In vertebrates the forelimbs and their supports are termed the *pectoral* appendages and pectoral girdle; and those of the hindlimbs are termed the *pelvic* appendages and pelvic girdle.

a. Pectoral Appendages and Girdle—The *clavicles* or collar bones help to hold the shoulder out away from the rib cage, and they serve as a jib from which the arm is hung; the clavicles are supported by muscles extending from the neck to the shoulder. The other important part of the pectoral girdle is the *scapula* or shoulder blade, which articulates by a *ball and socket joint* with the *humerus*, or upper arm bone. The humerus forms a *hinge joint* with the *ulna* of the forearm. Which bone forms the elbow? _____
The *radius*, adjacent to the ulna on the thumb side of the arm, has a wheel-like process. Demonstrate the function of this process by turning the wrist 180 degrees. The wrist consists of eight small *carpal bones*. Articulating with the carpal bones are five separate elongated *metacarpal bones*. Distal to the metacarpals are the *phalanges*.

How many phalanges are in the thumb? _____

How many are in the other fingers? _____

b. Pelvic Appendages and Girdle—In the pelvic girdle and leg, notice the relative length and strength of bones, as compared to those of the arm, related to the upright posture and bipedal mode of locomotion. The massive thigh bone, or *femur*, forms an obvious *ball and socket joint* with the pelvic girdle above, and articulates with the *tibia* of the shank below. Does the smaller shank bone, the *fibula*, permit the foot to rotate as the radius does the hand?

Move your hand down along the lateral side of your leg until you feel a prominent enlargement at the "ankle" region. This enlargement is called the lateral malleolus and is actually part of which bone? _____ On the skeleton examine the bones of the foot. Note the large heel bone (calcaneus). It plays a major role in transmitting the weight of the body to the ground. The calcaneus and the next six bones are collectively referred to as the *tarsal bones*. Attached to the tarsal bones are five separate elongated bones called the *metatarsals*. Note the arches formed by the tarsals and metatarsals. The distal bones of the foot are the *phalanges*. The "big" toe has two phalanges. How many phalanges are present in the other toes? _____

B. Muscle Action

Because muscles do their work by shortening (contracting), each end must be attached to some structure, permitting the contraction to draw the two points of attachment closer together. The relatively fixed attachment is, by convention, called the *origin,* and the muscle is said to arise or originate from this point. The more moveable attachment is called the *insertion,* and the muscle is said to insert at that point. The precise movement involved in any given case is called the *action* or the function of the muscle under consideration. In general, the action of muscles can be classified as flexion, extension, abduction, adduction or as modified combinations of these movements. With reference to the human body, practice the movements while reading over the terminology.

1. *Flexion*—the forearm is flexed onto the upper arm when the elbow is bent moving the hand upward toward the shoulder. The fingers are flexed when the hand is closed to make a fist. Flexion is the bending of a limb or limb part resulting in a decrease in the angle between the bones involved.

2. *Extension*—the forearm is extended away from the upper arm when the elbow is straightened so the hand moves away from the shoulder. The fingers are extended when the fist is opened. Hence, extension and flexion are opposite movements.

3. *Abduction*—the arm is abducted when it is moved in a lateral direction away from the median plane of the body, that is away from the sides. The fingers are abducted when they are spread apart.

4. *Adduction*—is the opposite of abduction. The arm is adducted when it is moved toward the median plane of the body, that is, lowered to the sides from the abducted position. The fingers are adducted when they are moved together from the spread position.

5. Allow your right arm to hang loose at your side with your palm facing forward (i.e., thumb on outside). Place your left hand on your upper right arm just below the shoulder. The main muscle you are touching is the *biceps brachii.*

   Now make a fist with your right hand and raise it as close to your right shoulder as you can. Do you detect a change in the biceps? With your arm raised, is the biceps in the relaxed or contracted state? If contraction of the biceps brings the arm up toward the shoulder, what is its action called? _____ Where do you think the origin of the biceps is located? _____ Where is its insertion? _____

6. Try and locate a muscle which has the opposite action of the biceps. Use Figures 21–7 and 21–8 to find its name.

7. Locate three more muscles on your body and identify the action, origin and insertion.
   a. gastrocnemius—
   b. deltoideus—
   c. sternocleidomastoideus—

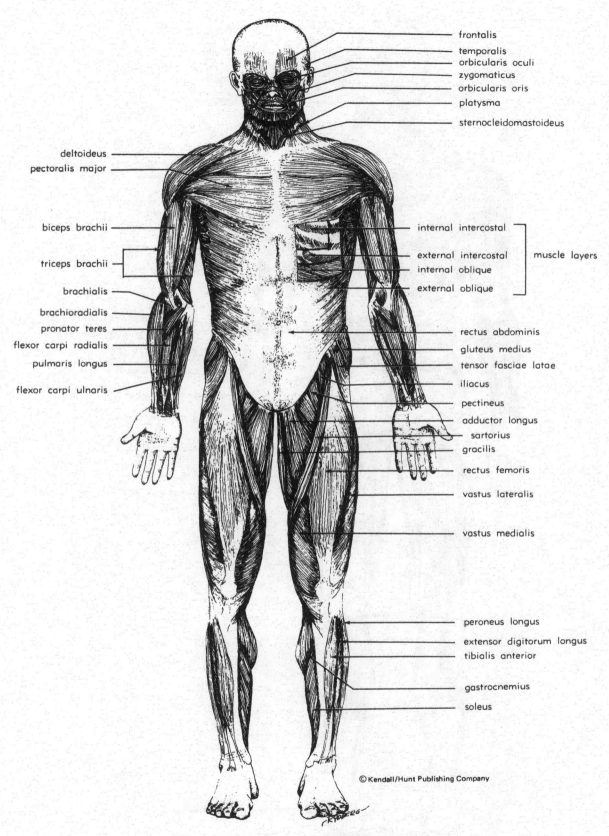

frontalis
temporalis
orbicularis oculi
zygomaticus
orbicularis oris
platysma
sternocleidomastoideus

deltoideus
pectoralis major

biceps brachii

triceps brachii

brachialis
brachioradialis
pronator teres
flexor carpi radialis
pulmaris longus

flexor carpi ulnaris

internal intercostal
external intercostal
internal oblique
external oblique

muscle layers

rectus abdominis
gluteus medius
tensor fasciae latae
iliacus
pectineus
adductor longus
sartorius
gracilis
rectus femoris
vastus lateralis

vastus medialis

peroneus longus
extensor digitorum longus
tibialis anterior

gastrocnemius
soleus

© Kendall/Hunt Publishing Company

Figure 21-7. Human, muscle system, anterior view.

241

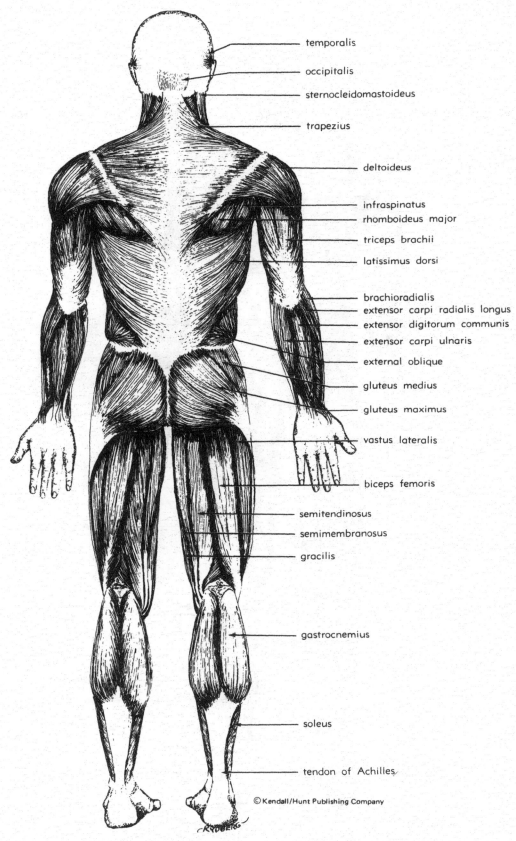

temporalis

occipitalis

sternocleidomastoideus

trapezius

deltoideus

infraspinatus
rhomboideus major
triceps brachii
latissimus dorsi

brachioradialis
extensor carpi radialis longus
extensor digitorum communis
extensor carpi ulnaris
external oblique
gluteus medius
gluteus maximus
vastus lateralis

biceps femoris

semitendinosus
semimembranosus
gracilis

gastrocnemius

soleus

tendon of Achilles

Figure 21–8. Human, muscle system, posterior view.

242

REVIEW QUESTIONS

1. Name the three basic portions of a neuron.

   a.

   b.

   c.

2. Diagram and label a simple reflex.

3. List the main components of the central nervous system.

4. Name the part of the brain responsible for:

   a. coordination of voluntary muscle action

   b. regulation of water balance

   c. imagination and creativity

5. Identify the specific division of the nervous system associated with each of the following:

   a. stimulation of skeletal muscle contraction

   b. control and coordination of skeletal muscles

   c. conveys impulses from receptors to the brain and spinal cord

   d. controls and coordinates actions involved in speaking

   e. conducts impulses that cause an increase or decrease in the heart rate.

6. List three functions of the skeletal system.

   a.

   b.

   c.

7. Indicate if the following bones are part of the axial skeleton or appendicular skeleton.

   a. skull

   b. ribs

   c. sacrum

   d. scapula

8. List five types of joints and cite an example of each.

    a.

    b.

    c.

    d.

    e.

9. List and explain four types of muscular action.

    a.

    b.

    c.

    d.

NERVOUS CONTROL: RECEPTORS

Objectives

Upon completion of this exercise you should be able to:

1. Identify the three basic factors common to all types of nervous control.
2. Define the term "stimulus" and give three examples of stimuli.
3. Distinguish between general and special sense receptors and give examples of each type.
4. Name the function of and identify from a microscopic slide or photomicrograph a Pacinian corpuscle and a taste bud.
5. Determine differences in touch sensitivity among various areas of the body.
6. Identify on specimens, models and diagrams and give the functions of the main parts of the vertebrate eye.
7. Calculate visual acuity if given appropriate data.
8. Identify on models and diagrams and give the functions of the main components of the outer, middle and inner ear regions.
9. Tell if a written description refers to accommodation, myopia, hyperopia, astigmatism, dominant eye, positive afterimage, negative afterimage, pink eye, glaucoma, or blind spot.
10. List the four basic taste sensations and locate the region(s) of the tongue where each sensation is most acute.
11. Answer all questions in the exercise:

I. Introduction

You are busy studying but become aware that a roast has been taken from the oven so you immediately close your book and rush into the kitchen for a taste.

Your dog is sleeping in the grass but suddenly leaps up and rushes to greet you just as your car comes around the corner.

Your car has been sitting outside overnight sometime in February. You get in it about 7:35 A.M. and sit on the vinyl seat. You immediately get up and place your notebook on the seat and then sit down again.

All of these situations have three basic things in common. Can you identify the three factors? The three basic factors each of the examples have in common are:

1. A detection of a change in the environment (a stimulus).
2. A transmission of a nervous signal (a nerve impulse).
3. A response to the stimulus (an effect).

For each situation, indicate the main stimulus and the main response.

	Stimulus	**Response**
Situation 1		
Situation 2		
Situation 3		

In situation 1, the odor of the roast is the stimulus. The molecules of the meat's juices diffuse into the air and eventually enter your nasal cavity. The presence of these molecules alters the environment surrounding the cells of your nasal epithelium. These specialized cells detect this change and cause a nervous signal to travel to your brain. As the impulse reaches specific areas in your brain, you "become aware" of the roast's aroma. Nervous signals now travel from the brain to various muscles which ultimately result in the closing of your book and the rushing into the kitchen.

In situation 2, the stimulus is the sound of your car as it is detected by the ear of the dog. The response is the awakening of the dog and his subsequent rush toward you.

In situation 3, specialized cells in your skin detect the coldness of the seat (cold is the stimulus). The response involves getting off of the seat.

The stimuli in the above examples included chemical (odor of roast), auditory (sound), and tactile (contact with cold seat). How many other types of stimuli can you list?

For each of these stimuli there is a highly specific receptor. A receptor may be the end of a single nerve cell or may be specialized cells in a highly complex structure such as the eye or ear.

II. General Sense Receptors

Receptors that are widely distributed throughout the body are classified as general sense receptors. These receptors are sensitive to pain, temperature, pressure, and touch.

A. *Identification of Receptors*

1. *Pain*—The most numerous receptors in the body are those for the sensation of pain. These receptors are simply the *free ends of nerve cells*. They respond to any type of intense stimulus.

2. *Temperature*—Within the skin are the ends of certain nerve cells that are specialized to detect changes in temperature. The receptors that are sensitive to cold temperatures are called the *end bulbs of Krause,* while the receptors sensitive to heat are called the *corpuscles of Ruffini.*

3. *Pressure*—Deep within the skin and around the internal organs (viscera) are specialized receptors surrounded by many layers of connective tissue that are stimulated by pressure. These are termed *Pacinian corpuscles.*

4. *Touch*—Hold your pen or pencil horizontally about two inches over the back of your hand or arm. Slowly move the instrument back and forth, but with each pass slowly lower it until it just touches the tip of the hairs on your arm. The movements of the hairs on your arm have stimulated certain specialized nerve ends located at the base of each hair. A little deeper within the skin are additional receptors for touch. These are called the *corpuscles of Meissner.*

It is important to note that these general sense receptors only serve to detect a stimulus. If the stimulus is strong enough, a nerve impulse will be generated and carried to a specific region of the brain. It is only when the impulse reaches the brain that the sensation may be felt.

Proceed to the demonstration area and identify the various general sense receptors. After viewing the demonstration, obtain a microscope and a prepared slide of a Pacinian corpuscle. Based on the previous description of the Pacinian corpuscle, identify and sketch a Pacinian corpuscle.

The Pacinian corpuscle is the receptor for _____ .

B. *Calculation of Touch Sensitivity Variation*

Are all parts of the body equal in their touch sensitivity? The sensitivity of an area is dependent on the position and number of receptors in that area. To calculate the touch sensitivity of various parts of your body, you will use the *two-point discrimination test.* It will be necessary for you to work with a partner for this experiment. The materials you will need are:

1. Two pins inserted in a piece of cardboard (or a two-pointed compass).

2. A millimeter ruler.

Have your partner close his or her eyes. The examiner will touch the parts of the body outlined in the touch sensitivity chart below with the points of two pins inserted in a piece of cardboard. Spread the pins apart and touch with the double points.

Ask the subject if he or she feels one or two points touching. If the answer is "two" when two are touching, reduce the distance and repeat. Continue reducing the distance until only one point is perceived. Be sure that both points are applied with nearly equal pressure. Measure, in millimeters, the distance between the two points—at the two point position identified by the subject as one point. Record the distance on the chart. Repeat for each part of the body and then exchange places with your partner.

Touch Sensitivity
Distance Between Points of Pins

Area	Subject #1	Subject #2	Class Average	Area	Subject #1	Subject #2	Class Average
Back of Hand				Forehead			
Palm				Nose			
Finger Tip				Back of Neck			
Forearm				Front of Neck			
Cheek				Lip			

After completing the entire chart, record your data on the board or on the class chart at the demonstration area. Compute the class average and compare this with your own results.

Which area had the *smallest* distance between the two points? _____

This is the area that contains the greatest number of touch receptors and therefore should be the most sensitive. Rank all areas tested in terms of their relative sensitivities.

III. Special Sense Receptors

If general sense receptors are so named because they are widely distributed throughout the body, what do you think is the basis for the special sense receptors? These are receptors that are located in *specific* areas of the head. Try to name four of these special receptors and the type of stimulus to which they respond.

Receptors which respond to light energy are located in the *eye*. Receptors for balance and hearing are located in the *ear*. Receptors for the chemical senses of smell and taste are located in the *nasal cavity* and *tongue*, respectively.

A. *Anatomy of the Eye*

1. *External Anatomy.* Study your own eyes in a mirror (or your partner's if no mirror is handy).

   a. Observe each of the following structures and list the function of each.

      (1) eyebrows

      (2) upper and lower eyelids

      (3) upper and lower eyelashes

   All of the above structures *aid* in protecting the eye.

   b. Another structure which functions in protection is a transparent mucous membrane, the *conjunctiva*. The conjunctiva lines the inner surface of each eyelid and continues over a portion of the eyeball itself. It contains many goblet cells which secrete a mucus material.

   "Pink eye" is the common name for the condition existing when the conjunctiva is infected. The conjunctiva is normally protected from infection by a continuous flow of tears produced by the *lacrimal gland*. The tears bathe the anterior surface of the eyeball, wash away particles that might "enter" the eye and contain an enzyme that destroys certain microorganisms. The tears flow from the upper lateral surface of the eyes across to a pinkish triangular mass of tissue, the *lacrimal caruncula*, located in the medial angle between the eyelids (nearest the nose). From here, the tears normally drain into the nasal cavity through the *nasolacrimal duct*. In certain instances the tears may overflow and run down the cheek. This may be due to hypersecretion by the lacrimal gland brought on by emotional crisis, irritation of the conjunctiva, etc. or to a blockage of the nasolacrimal duct brought on by a cold or flu.

   c. The "white" area of the eye is a dense fibrous connective tissue layer called the *sclera*. It functions in protection and support. Observe your partner's eye from the side. Notice that the sclera is modified at the anterior region of the eye. The modified area is transparent and is called the *cornea*. The cornea bends light rays and allows them to enter the eyeball.

   d. By looking through the transparent cornea you will be able to see a small dark hole in the center of the eye. The opening is the *pupil* and is the entrance into the eye for light rays. Surrounding the pupil is a pigmented ring, the *iris*, which serves to regulate the amount of light entering the eye.

2. *Internal Anatomy.* Utilize the eye specimens or models for this section.

   a. The eye is surrounded by three basic layers. The first of these, examined in section 1c, functioned to protect and support the eye and is called the _____ .

   b. Internal to the sclera is the middle layer or *choroid*. It contains many pigments and blood vessels. The anterior portion of the choroid has been modified to form the *ciliary body* and *iris*. Attached to the ciliary bodies are the *suspensory ligaments* which support the *lens* of the eye. Contraction of the muscles of the ciliary body results in altering the lens shape for viewing near objects.

248

c. The innermost layer of the eye is the *retina* and contains the photoreceptor cells, *cones* and *rods,* that are sensitive to light stimuli. At the rear of the eye are two areas:

 (1) One area at the center of the retina contains only cones that are sensitive to color and detail. This area, the *fovea centralis,* is the region of most acute vision.

 (2) Medial to the fovea centralis is an area devoid of receptor cells. This area, the *blind spot,* is where the nerve processes come together to form the *optic nerve.* The optic nerve carries the nerve impulse to the brain, where it is finally interpreted.

d. The space between the cornea and the front of the iris is the *anterior chamber.* The space behind the iris and in front of the lens is the *posterior chamber.* Fluid filters from the capillaries in the ciliary body and fills both chambers. This fluid, *aqueous humor,* helps maintain the proper shape of the eyeball. Excess fluid drains back into the blood through a small canal in the anterior chamber. If the drainage is impaired, the pressure within the eye builds up resulting in abnormal vision and much pain. This condition is called *glaucoma.*

e. Posterior to the lens is a space filled with a jellylike material, the *vitreous body.* It aids in maintaining the shape of the eye and in supporting the retina.

f. Locate and label on Figure 22–1 all the italicized terms mentioned in Section A.

B. *Sensory Reception by the Eye*
 1. *Visual acuity*—Stand with your toes at a mark located 20 feet from a visual acuity chart. Cover one eye, and try to read the indicated lines on the chart. The numbers by each line indicate the distance at which the normal eye should be able to identify the designated letters or symbols. If at 20 feet you can read letters that individuals with normal eyesight can read at that distance, your vision is 20/20 (normal). If you can, while standing 20 feet from the chart, read only the larger letters which normally one should be able to read at 40 feet, your vision is 20/40 or nearsighted. In *myopia* or nearsightedness, the eyeball may be too long or the refractory (light bending) power of the lens too great, making images fall in front of the retina. Wearing concave lenses in spectacles corrects the optical error. On the other hand, if the eyeball is too short or the power of the lens too low, images fall "behind" the retina (apparent image, that is), in the condition known as *hyperopia* or farsightedness. A convex lens is needed in the spectacle to correct this condition.

 What is the visual acuity of your right eye? _____ Your left eye? _____
 If you wear glasses, take the test a second time wearing them, and note whether they correct the condition.

 Check also for *astigmatism* if your chart is equipped with a sunburst of black lines. If the lines on one side appear indistinct when the eye is focused on the lines on the other side, the eye is said to show astigmatism. This is usually due to differences in vertical and horizontal curvature of the cornea, or sometimes the lens. Hence, if the eye is accommodated for vertical lines, horizontal lines will be indistinct, or vice versa. In correcting this condition, lenses with the same curvatures as the cornea are made, but they are held so the greater curvature of the lens corresponds to the lesser curvature of the cornea, and the lesser curvature of the lens to the greater curvature of the cornea.

 2. *Accommodation and pupillary reflex*—Normal relaxed eyes see a distant object very well. The lenses are pulled rather flat by ligaments and images focus clearly on the retina. If the eyes look at a nearby object, such as the printed matter in this exercise, they must *accommodate.* Light rays must be bent more in order to be focused. A set of circular muscles surrounding the lens contracts causing the elastic lens to bulge in the middle, shortening the focal length and bringing the image into sharp focus. *Accommodation* is the term applied to the ability of the lens to change shape in order to maintain good sharp vision.

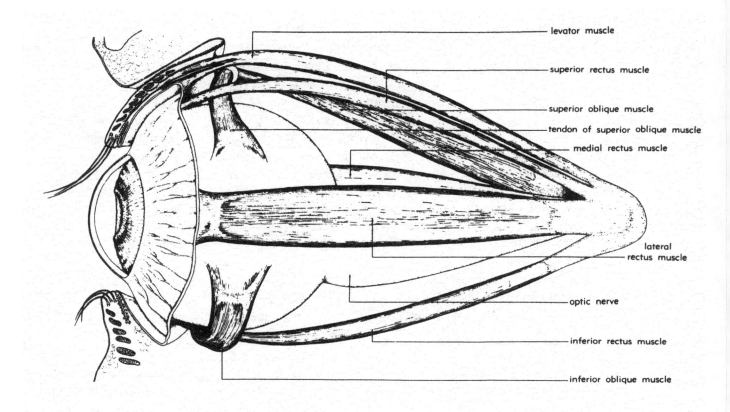

levator muscle

superior rectus muscle

superior oblique muscle

tendon of superior oblique muscle

medial rectus muscle

lateral rectus muscle

optic nerve

inferior rectus muscle

inferior oblique muscle

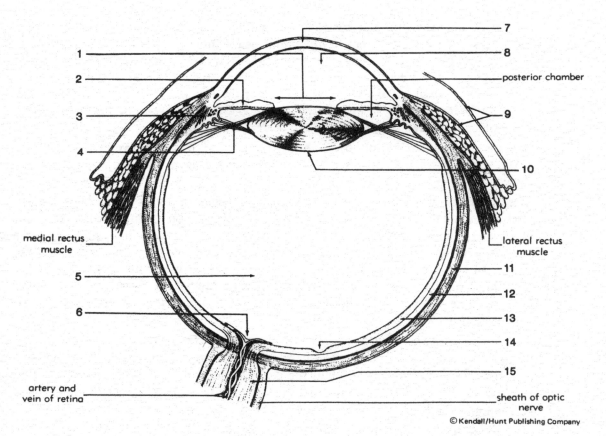

1

2

3

4

5

6

7

8

posterior chamber

9

10

11

12

13

14

15

medial rectus muscle

lateral rectus muscle

artery and vein of retina

sheath of optic nerve

Figure 22-1. Human, eye.

a. Have someone look at a distant object. Then have him (her) look at the print on this page about 30 cm from his (her) nose. What happens to the pupils?

_____

The pupils should constrict (become smaller).

b. Hold the laboratory book at arms length in front of you. Close one eye. Focus on the page number. Slowly move the book closer to you until the page number passes out of focus. Measure the distance in centimeters between the page and your eye.

Repeat the process using your other eye. If you wear glasses, perform the exercise first with your glasses then without them. Record all measurements below:

Distance at Which
Number Passes Out of Focus

Right eye _____

Left eye _____

The *shorter* the distance between your eye and the page, the *greater the elasticity* of your lens. An average distance for a young adult is about ten centimeters. As an individual ages, the elasticity of the lens decreases.

c. Have someone cover both eyes for a short while. What happens to the pupils when the eyes are uncovered? _____
The pupils should constrict.

d. Shine a light (pen light) into one eye. What happens to both pupils? _____
Both should constrict.

The constriction and dilation of the pupil is controlled by what structure?

_____ (Check your answer by rereading Section A.1.d.).

3. *Blind spot*—On a plain sheet of paper draw a black (dark) circle about 1 cm in diameter. About 10 cm to the right of the circle, draw a heavy black cross approximately the same size. Hold the sheet about 50 cm away from your face and cover your left eye with the free hand. Direct your gaze on the circle and *slowly* move the paper nearer. At a certain distance, the cross will disappear from view. Light from the cross is now falling on your blind spot.

Move the sheet still nearer, and note what happens: Does the cross reappear? _____

You may wish to check the other eye, but turn the paper around so that the cross is to the left.

Anatomically, the blind spot represents what? _____

4. *Binocular vision*—Place your index fingers tip to tip about 25 cm in front of you, on a level with your eyes. Now, looking past your fingers, focus on the person directly across the table from you.

How do the fingers appear?_____

Now, look back at the fingers. Is the person in focus? _____

How many images of the individual do there appear to be? _____

Initially when your eyes were focused on the individual, your fingers should have appeared out of focus. When your eyes converged to view the fingers, the individual appeared out of focus. The individual appeared as two overlapping images due to the fact that when an image is not in sharp focus, it falls upon *different* locations on the two retinas.

5. *Dominant eye*—Select an object (not too large) at the front of the room. Point at it with your arm and index finger extended. Aim the finger with both eyes open. Now, without changing the position of either head or arm, close first one and then the other eye. With which eye does the object stay in line? This is the dominant eye.

6. *Afterimages*—Images that remain on the retina after the source of the image has been removed are called afterimages. A positive afterimage is identical with the original subject. Negative afterimages have colors which differ from the original.

   a. *Positive afterimages:* Focus your eyes on one spot on the light of a gooseneck lamp about one meter away for one minute, and then close your eyes or turn them to a dark surface such as the desk top. An image of light should slowly float in view, remain bright for a while and then disappear.

   Positive afterimages are thought to be caused by the photochemical activities of the retinal receptors which outlast the stimuli that produce them.

   b. *Negative afterimages:* Stare at the lamp again for one minute and then look at a sheet of white paper. The images should now be a dark gray or black. The negative image of white light, therefore, is black. Use individual pieces of colored paper (red, yellow, blue) at desk level; stare at them for one minute and then look at a sheet of white paper.

   Negative afterimages are thought to be caused in part by momentary fatigue of specific receptors. When the eyes are directed to a white surface, only those receptor cells that had not previously been stimulated respond and a complementary color is perceived. However, there are many complexities, and this does not completely explain the phenomenon.

7. *Defects in color vision*—Most people are able to distinguish between blues, reds, yellows and greens with little difficulty. In a few rare cases, a person is found whose retina is totally insensitive to colors and everything appears black and white or shades of gray. Partial color blindness is more common, in which receptors for specific colors (often red or green) may be absent from the retina or the perception of these colors may simply be below normal.

   Many people suffer from some form of color blindness and are unaware of it. Even though their color perception is defective, they still are able to make some sort of distinction between colors. They have learned to give color names to familiar objects based on characteristic brightness and are able to distinguish many colors by a recognition of slight brightness difference. A colorblind testing book may be available to use to test your own color vision.

8. *Optical Illusion*—The human eye can distinguish size, depth, and distance as well as color and form. But seeing is not always believing, for the eye can be tricked. For each figure below identify the longer line. Check your answers by measuring each line.

C. *Anatomy of the Ear*

Using the descriptions below, locate each italicized structure on an ear model and then label Figure 22-2.

The ear can be divided into three parts: an outer ear, a middle ear and an inner ear.

1. *The outer ear* consists of the ear flap, or *pinna*, and the *auditory canal*. The *tympanic membrane* (ear drum), a flexible membrane, separates the outer and middle ear cavities.

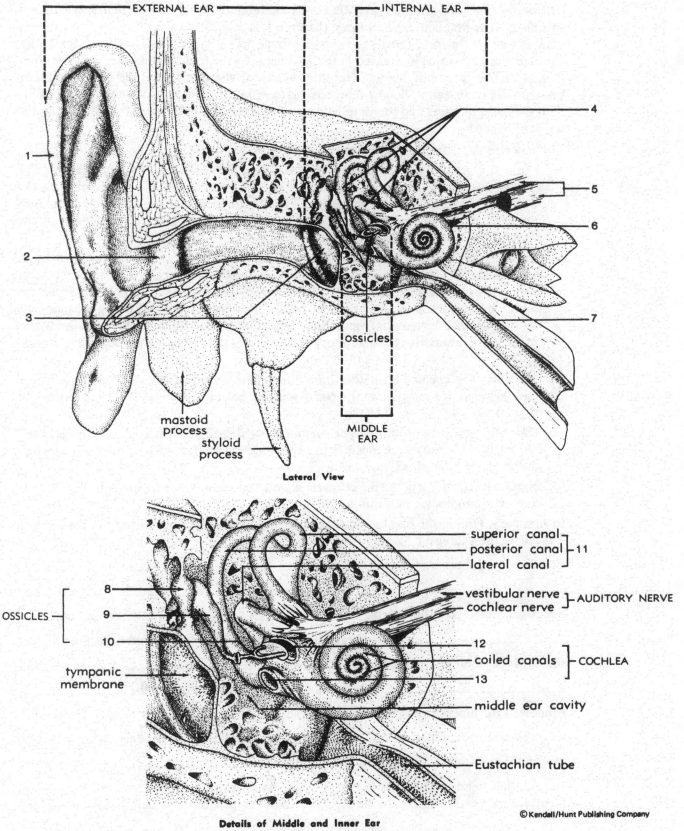

EXTERNAL EAR

INTERNAL EAR

4

5

1

6

2

3

7

ossicles

mastoid
process

styloid
process

MIDDLE
EAR

Lateral View

superior canal
posterior canal
lateral canal
11

OSSICLES

8

vestibular nerve
cochlear nerve
AUDITORY NERVE

9

10

12

coiled canals
COCHLEA

tympanic
membrane

13

middle ear cavity

Eustachian tube

© Kendall/Hunt Publishing Company

Details of Middle and Inner Ear

Figure 22-2. Human ear, general anatomy.

2. The *middle ear* is an air filled cavity in bone containing three tiny bones (*ossicles*) arranged in a chain. The first bone, the *malleus,* (hammer), is connected to the tympanic membrane. Next in line are the *incus* (anvil) and *stapes* (stirrup). The stapes is attached to a membrane, the *oval window,* which is situated in the wall between the middle and inner ear. The three bones function in amplifying and transmitting sound vibrations from the tympanic membrane to the inner ear. A hollow tube, the *eustachian tube,* connects the middle ear cavity with the nasal pharynx and serves to help equalize the pressure between the middle ear and the atmosphere.

3. The *inner ear* consists of several fluid filled chambers.

   a. The first chamber, the *cochlea,* resembles a snail shell in that it is tightly coiled. The cochlea attaches to the stapes by means of the oval window. As the stapes vibrates, the oval window is pushed in causing the fluid within the cochlea to move. The pressure thus generated within the inner ear is relieved by the bulging out of a second membrane, the *round window* located just below the oval window.

   Inside the cochlea are three fluid filled spaces. The middle space, the *cochlear duct,* contains the *organ of Corti* on its *basilar membrane.* The organ of Corti consists of several rows of receptor cells (hair cells) covered by a rooflike *tectorial membrane.* As fluid in the cochlea is set in motion, the basilar membrane vibrates causing the hair cells to brush against the tectorial membrane. This stimulation initiates nerve impulses which travel to the brain along the auditory nerve. Study the cochlea cross section in Figure 22–3.

   b. Just above the cochlea is another fluid-filled chamber called the *vestibule.* It contains hair cells that are sensitive to the position of the head in space (as it relates to the pull of gravity).

   c. Above the vestibule are three *semicircular canals,* each at right angles to one another. Hair cells in the canals are sensitive to any changes in motion, such as starting, stopping, changing speed or direction.

   Impulses originating in the semicircular canals and vestibule travel to the brain and thus help maintain body equilibrium.

4. In summary: (Fill in the blanks and then check your answers by reviewing the appropriate sections of the exercise.)

   The three major parts of the ear are _____ ,
   _____ , and _____ .

   Sound vibrations pass through the _____ canal of the outer ear and cause the
   _____ to vibrate. These vibrations are transferred across the middle
   ear by three small bones, the _____ , _____ , and
   _____ . Vibrations of the stapes push against the _____
   window of the inner ear and set the fluid in the _____ in motion. To equalize
   the pressure in the inner ear the _____ window bulges out. The
   moving fluid in the cochlea causes the _____ membrane to
   vibrate, thus bending the hairs of receptor cells against the _____
   membrane. As the hairs are bent, an impulse is generated which passes along the
   _____ nerve to the brain.

   Two additional structures are present in the inner ear and serve to help maintain body equilibrium. These are _____ and the three _____ .

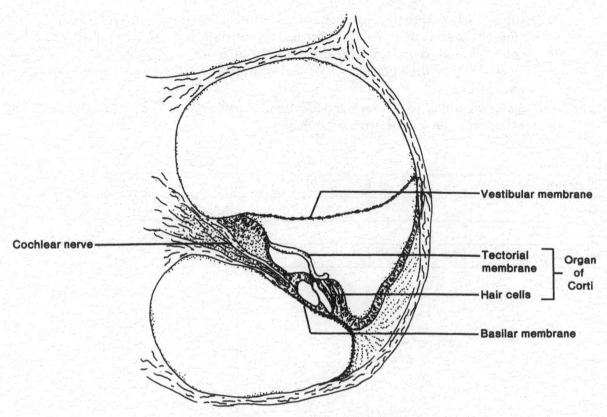

Figure 22-3. Cochlea, cross section.

D. *Anatomy of Taste and Olfactory Receptors*

1. *Taste Receptors*—Receptors for taste are found in the mucous membrane covering the tongue, but to a lesser extent may be located elsewhere. The *taste buds,* which contain the special receptor cells, are located in the sides of little pits in the surface and sides of the tongue. Using the microscope, identify the taste buds in a cross section of a mammalian tongue. Sketch a taste bud.

2. *Olfactory Receptors*—In reality, many of the sensations we call taste are really due to the sense of smell. High in the nasal cavity on the underside of the cranium or brain case is a very small patch of specialized epithelium, the olfactory or smell receptor area. Odoriferous substances emit particles which are usually in the gaseous form. Molecules of the odorous material dissolve in the fluid covering the smell receptor cells thereby setting off nerve impulses which the brain interprets as sensations of odor or smell.

E. *Taste and Odor Tests*

1. There are only four basic taste sensations: sweet, bitter, sour and salty. All other taste "flavors" are combinations of these four and may include smell. There are two major cranial nerves that control the taste sensations of the tongue. The *facial nerve* (cranial nerve VII) receives impulses from the anterior two thirds of the tongue and the *glossopharyngeal nerve* (cranial nerve IX) from the posterior one third.

2. Have one person per table chew a stick of peppermint gum while holding his (her) nose. Is the sensation of peppermint due to taste or aroma?

3. Map the four taste sensations for the same individual in the following manner. Place approximately one ml of 10% salt (NaCl) solution in a small container. Make an applicator by wrapping a bit of cotton on a toothpick. Dip the applicator into the solution. Rinse your mouth with water to remove any taste present. Rest the soaked applicator briefly on the

255

(1) tip, (2) front center, (3) back center and (4) side of the tongue. Avoid applying excess solution. Note and record on the chart below, where the salty taste is intense (++), less intense (+), and absent (0). Repeat the test for each of the following solutions: 5% sugar, 1% acetic acid and 0.1% quinine sulfate. Between applications rinse your mouth with a small amount of water.

Make a diagram of your tongue and identify the taste regions. Attempt to associate a taste sensation with the cranial nerve controlling it.

Taste	Tip	Front Center	Back Center	Sides
Salty (Salt)				
Sweet (Sugar)				
Sour (Acetic Acid)				
Bitter (Quinine Sulfate)				

REVIEW QUESTIONS

1. A detectable change in the environment that elicits a response is called a _____

2. List four types of general receptors.

   a.

   b.

   c.

   d.

3. Which of the four is represented by just free nerve endings? _____

4. The sclera of the eye is commonly referred to as _____

5. List the three main coats of the eye.

   a.

   b.

   c.

6. Which eye layer contains the light sensitive photoreceptor cells? _____

7. The photoreceptor cells involved in forming sharp colored images are called _____

8. The organ of Corti is found in what component of the inner ear? _____

9. What is the function of:

   a. malleus, incus, and stapes

   b. semicircular canals

10. What is meant by the term "myopia?" _____

_____

11. The ability of the lens to change shape in order to maintain good sharp vision is called

_____

12. What is the difference between a positive and negative afterimage? _____

_____

_____

13. What does the blind spot represent in terms of the anatomy of the eye? _____

_____

14. Which area of the tongue is most sensitive to bitter taste? _____

15. What is the function of the eustachian tube? _____

16. Diagram an eye and label: cornea, anterior chamber, lens, suspensory ligaments, ciliary body, sclera, choroid, retina, vitreous body, fovea centralis, blind spot and optic nerve.

PLANT BODY AND LEAF STRUCTURE AND FUNCTION

Objectives

Upon completion of this exercise you should be able to:

1. List the principal vegetative and reproductive parts of flowering plants.
2. Distinguish between the terms:
   a. woody and herbaceous
   b. annual, biennial and perennial.
3. Recognize microscopically, the cell and tissue types listed below:
   a. meristem
   b. epidermis
   c. parenchyma
   d. chlorenchyma
   e. collenchyma
   f. sclerenchyma (fibers, stone cells)
   g. endodermis
   h. xylem (tracheids, vessel elements, fibers and parenchyma)
   i. phloem (companion cells, sieve tube elements, fibers and parenchyma).
4. Associate each cell and tissue type listed in objective 3 with a written description.
5. Define or identify:
   a. pit
   b. lumen
   c. Casparian strip
   d. ray
   e. sieve plate.
6. List and identify the main parts of a simple and a compound leaf.
7. Identify the main tissues in a leaf cross section.
8. Contrast monocot and dicot leaves.
9. Answer all review questions in the exercise.

I. Introduction

Within the plant kingdom there is tremendous variety. Plants range in size and complexity from microscopic forms to the giant sequoias and redwoods over 300 feet tall. The plants with which most humans are familiar are the angiosperms or flowering plants. This and two other exercises will treat the principal structural features and processes of flowering plants.

Over 250,000 species of flowering plants are known. These are divided into two classes: *dicots* and *monocots*. Dicots have two cotyledons ("seed leaves") associated with the embryo in a seed. Important or familiar dicots include beans (green, yellow, snap, pole, kidney, etc.), potatoes, peppers, tomatoes, carrots, lettuce, celery, cucumbers, cauliflower, sugar beets, apples, *Coleus,* geraniums, roses, oaks, maples, dogwoods, peanuts, walnuts, mistletoe, tobacco, etc.

Monocots have one cotyledon. Important or familiar monocots include corn, rice, wheat, rye, barley, oats, other grasses, lilies, sugar cane, palms, orchids, pineapple, bananas, etc.

II. Importance of Flowering Plants

Which of the above did you eat yesterday? Did you eat bread? Salad? Potatoes? Rice? Peanut butter? Steak? Groundbeef? Bread comes from the seeds of wheat and/or rye plants. The common ingredients of salads are familiar dicots. Peanut butter is derived from peanuts. Steak and groundbeef are derived from cattle that eat various grasses. In short, humans are dependent on flowering plants for the food they eat and the oxygen they breathe. Mankind also uses flowering plants for indoor and outdoor landscaping, preparing flower arrangements, spices, dyes, medicines, beverages, and as fibers.

III. Organs of Flowering Plants

The principal organs of flowering plants are:

A. *Vegetative parts*—roots and shoots (stems and leaves). They are involved in the day to day existence of individual flowering plants. By definition each of these parts contains vascular or conducting tissues including xylem which conducts water and phloem which conducts food.

B. *Reproductive parts*—flowers, fruits, and seeds. They are involved with the sexual reproduction, survival and dispersal of flowering plants.

At this time obtain and examine the vegetative parts of representative angiosperms (see Figure 23–1.)

Flowering plants are either *woody* or *herbaceous*. The hard woody tissues provide support for the larger bodies of trees and shrubs. Herbaceous (nonwoody) plants have soft tissues and are relatively short lived. Many of the familiar weeds are herbaceous.

Flowering plants can be described in terms of their life cycle. An *annual* is a short-lived plant, the entire life cycle of which, from seed germination to seed production, takes place within one growing season. Rye grass, beans and peas are annuals.

Biennials have a life cycle that is spread over two growing seasons. Flowers and seeds are normally produced during the second growing season. Carrots and celery are examples.

Perennials are plants that live and grow for more than two years. Oaks and maples are examples of woody perennials. Lilies are annuals above ground, perennial below ground.

IV. Plant Cells and Tissues

There are various types of cells and tissues that compose the plant body. Plant cells are characterized by the presence of a cell wall of cellulose around their protoplasts. Cells of soft plant parts have thin *primary* walls. Cells of fibrous and woody tissues form a *secondary* wall inside the primary wall adjacent to the protoplast.

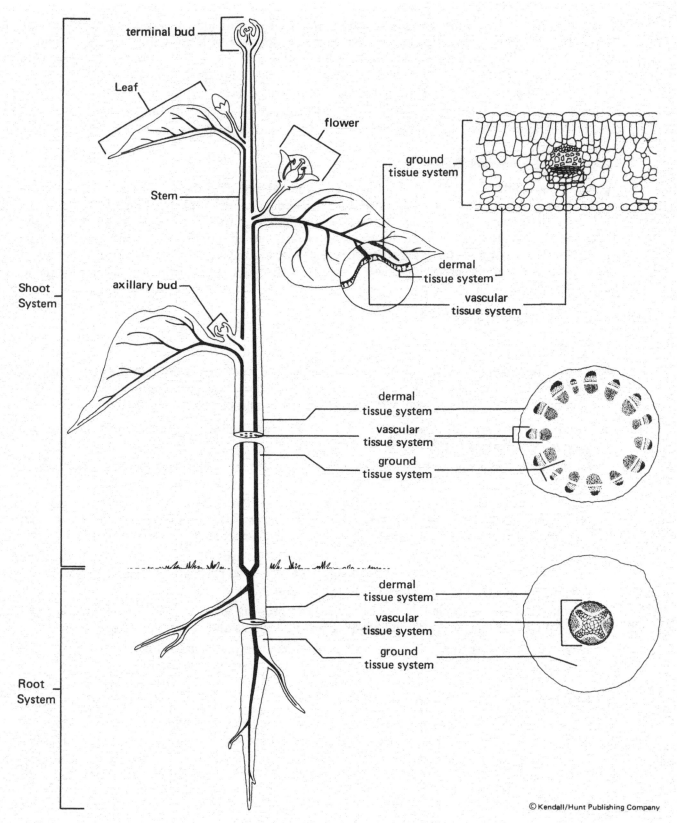

terminal bud

Leaf

flower

Stem

ground
tissue system

dermal
tissue system

vascular
tissue system

Shoot
System

axillary bud

dermal
tissue system

vascular
tissue system

ground
tissue system

dermal
tissue system

vascular
tissue system

ground
tissue system

Root
System

© Kendall/Hunt Publishing Company

Figure 23-1. Distribution of tissue systems in the seed plant.

A. *Meristematic Cells* (Figure 23–2A).

A region composed of cells that remain capable of division is called a *meristem*. Meristematic cells form all other cells. Obtain a prepared slide of a longitudinal section (l.s.) of an onion (*Allium*) root tip. The meristem in a root is located near the end or apex of the root. It is therefore an *apical meristem*. Covering the apical meristem of the root is a collection of nondividing cells called the *root cap*. Locate the cap. What is its function? _____

_____

The soil environment of most roots is quite abrasive. Root cap cells protect the apical meristem as the root grows through the soil.

Find the apical meristem located behind the root cap. Both dividing (cells in which chromosomes are visible) and nondividing cells can be seen in this region. Examine a nondividing cell. Is the wall thick or thin? _____ What proportion of the volume of the cell is occupied by the nucleus? _____

The cells of apical meristems are generally uniform in diameter and thin-walled, with a nucleus that is large in proportion to the volume of the cell. All other cell types are derived from meristematic cells by the processes of *cell division, cell enlargement* (elongation) and *cell differentiation* and are classified as dermal, ground or vascular tissue.

B. *Dermal Tissue*

Epidermis (Figure 23–2B).

Epidermal cells typically appear brick shaped when viewed from the side. They form the outer protective covering of leaves, young stems, roots, and in fact most young plant parts. Recall that you viewed and diagrammed onion leaf epidermal cells in surface view in the exercise on Cell Structure and Cell Division. You may want to review your diagram and notes.

C. *Ground Tissue*

1. Parenchyma (Figure 23–2D).

Parenchyma cells are living succulent or juicy cells having thin primary cell walls. They tend to be of equal diameter in any direction although they may be somewhat elongate. Intercellular spaces may be seen where adjacent cells do not meet.

Parenchyma functions in:

a. water storage in succulent parts of desert and other plants.

b. food storage as in carrots, potatoes, tomatoes, apples, etc.

c. photosynthesis as in leaf mesophyll.

d. parenchyma cells are also capable of dedifferentiating and becoming meristematic. This is important to a plant in the formation of a secondary body, to be discussed later, and in healing wounds.

Parenchyma cells are present in all plant parts. Specific locations include:

a. the mesophyll of a leaf.

b. cortex and pith in roots and stems.

c. xylem and phloem consist in part of parenchyma.

Recall having viewed parenchyma cells in a wet mount of potato. You may want to review your diagrams and notes. Alternatively, prepare a wet mount of potato tissue or look at the parenchyma in the center of a sunflower (*Helianthus*) stem cross section.

2. Chlorenchyma (Figure 23–2C).

Parenchyma cells that contain chloroplasts are sometimes called chlorenchyma.

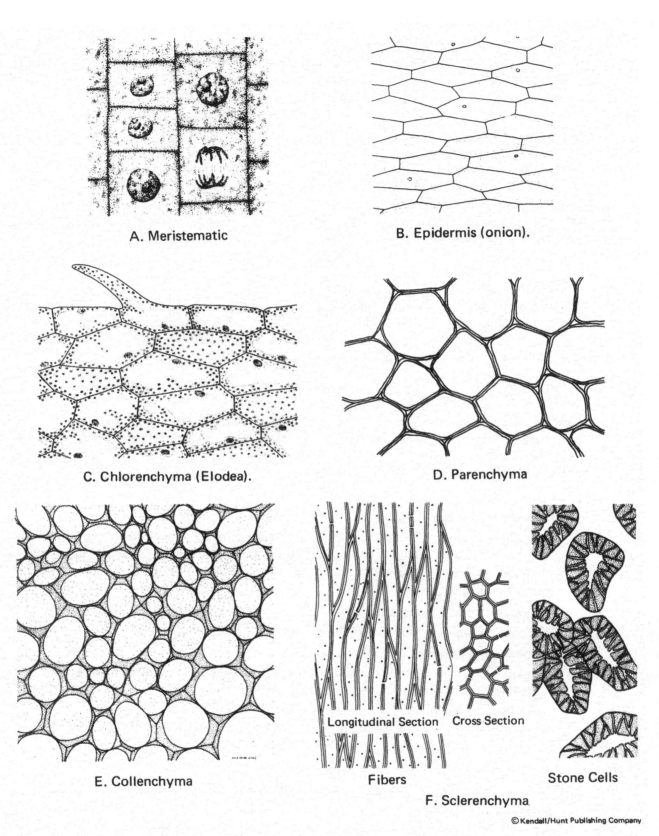

A. Meristematic

B. Epidermis (onion).

C. Chlorenchyma (Elodea).

D. Parenchyma

E. Collenchyma

Longitudinal Section   Cross Section

Fibers

Stone Cells

F. Sclerenchyma

Figure 23-2. Representative cells of the simple plant tissues.

3. Collenchyma (Figure 23–2E).

Collenchyma consists of living cells with unevenly thickened primary walls. In one type the primary walls are thickened at the corners obliterating the intercellular spaces seen in parenchyma. It is usually found in young actively elongating plant parts (stems and petioles) where it functions in support. Obtain a slide of a cross-section (c.s. or x.s.) of a *Cucurbita* stem (pumpkins, squashes, and cucumbers are cucurbits). Locate a region containing collenchyma cells just beneath the epidermis.

What shape are the epidermal cells? _____

Are intercellular spaces evident in collenchyma? _____

Now observe the same area in a *Cucurbita* stem l.s. Locate the epidermis and collenchyma.

What shape are the epidermal cells? _____ Are intercellular spaces evident

in collenchyma? _____

4. Sclerenchyma (Figure 23–2F).

Sclerenchyma cells possess thick, pitted, lignified primary and secondary cell walls. They differentiate from parenchyma-like cells or directly from meristems. The secondary wall is formed only after the cell has attained its final size and shape. Eventually, the cell dies. Such cells provide support in older plant parts that are not actively growing in length. Areas with little or no secondary thickening may be evident in the cell wall. These are termed pits.

  a. Fibers.

    Fibers are long, spindle-shaped cells. Obtain a slide of macerated ash wood (*Fraxinus*

    americana mac.). Sketch a fiber cell below. Can you see any pits? _____

    The pits may appear as unstained areas in the wall. Your drawing of a fiber should show a long cell tapered at both ends. The cell wall is thick and the cell interior (lumen) is narrow.

  b. Using the *Cucurbita* stem slides (c.s. and l.s.) examined previously, locate the cells just inside the collenchyma. These cells represent sclerified parenchyma.

    Is the wall thick? _____

    Are the cells short or elongated? _____

  c. Stone cells.

    Stone cells are somewhat rounded sclerenchyma cells responsible for the gritty nature of pears. Using a razor blade scrape some cells from the inside of a pear and prepare a wet mount. If the preparation is too thick for observation prepare a squash as follows:

    1) Put the coverslipped preparation on the lab table.

    2) Place a paper towel over the coverslip.

    3) Apply pressure in a downward direction with your thumb.

    4) Observe and sketch a stone cell.

Is the wall thin or thick? _____

Is the lumen narrow or wide? _____

Are pits evident? _____

Stone cells have a thick, pitted wall and a narrow lumen.

5. Endodermis

The endodermis is a layer of cells between conducting and storage and/or processing cells in a root and sometimes in a stem. The cells are sometimes surrounded by a suberized band called a *Casparian strip*. Its presence is thought to cause all water and minerals entering the vascular tissue to pass through the living material of endodermal cells. Obtain a prepared slide of a buttercup (*Ranunculus*) root c.s.

What tissue forms the outer covering of young roots? _____

Beneath the outer epidermis is a region, the *cortex,* consisting of parenchyma. Note the red stained material between the sidewalls of adjacent endodermal cells. What is this band of

material called? _____ Check your answer above.

D. *Vascular* (*conducting*) *Tissues*

1. Xylem.

Xylem is a complex tissue consisting of parenchyma, fibers and conducting cells (tracheids and/or vessels). Conducting cells have both primary and secondary walls of cellulose impregnated with lignin. The secondary wall of conducting cells is laid down in the form of rings, spiral bands, scalariform thickening or appears pitted. See Figure 23–3A. Water and mineral transport occurs in the lumen of dead conducting cells. Because of their thick walls, conducting cells also provide support.

a. Tracheids.

Tracheids are long overlapping cells with pointed ends. Water moves upward from cell to cell through thin areas of the walls called *circular bordered pits*. Obtain a slide of macerated pine wood (*Pinus australis* mac.). Sketch a tracheid.

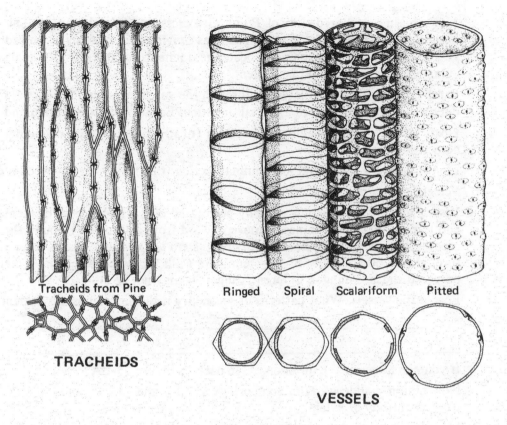

Tracheids from Pine

TRACHEIDS

Ringed   Spiral   Scalariform   Pitted

VESSELS

A. Xylem

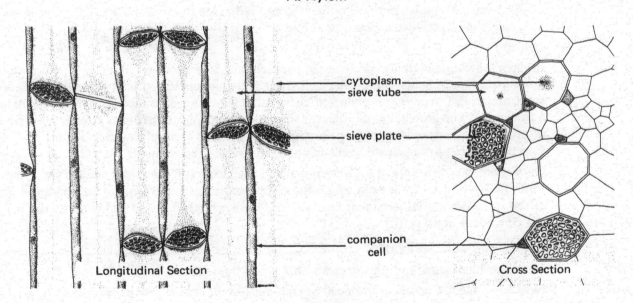

cytoplasm
sieve tube

sieve plate

companion cell

Longitudinal Section

Cross Section

B. Phloem

© Kendall/Hunt Publishing Company

Figure 23-3.

Now obtain a prepared slide of *Pinus* with cross, radial and tangential stem sections. Lengthwise stem sections are referred to as *longitudinal* sections. Both radial and tangential sections are longitudinal. Locate the tracheids on either longitudinal section. Circular bordered pits should be evident.

Parenchyma cells extend laterally through the vascular tissue forming *rays*. Rays conduct water and minerals in a lateral direction. Locate the cross section and find a ray.

Next, locate a radial longitudinal section (r.l.s.). Rays are seen extending lengthwise across the section.

Finally, locate a tangential longitudinal section (t.l.s.). Rays are viewed on end.

b. Vessels.

Vessels are tubes consisting of many perforated cells, called vessel elements, arranged end to end. Water and minerals move upward from cell to cell through the perforated areas. Again look at a slide of a *Cucurbita* stem l.s. Vascular tissues extend lengthwise through the other tissues of the stem. The walls of vessels are usually stained red in prepared slides. Younger, differentiating cells may not be stained.

Locate the vessels. What patterns of secondary wall thickenings are evident?

_____

_____

If you can't answer the question, refer back to Figure 23–3A.

Now obtain a slide of *Tilia* composite c.s., t.l.s. and r.l.s. Each of the three sections consists entirely of xylem.

Identify rays, vessels and fibers in both cross and longitudinal sections.

Can you distinguish between tangential and radial longitudinal sections of *Tilia?* If not, refer back to the discussion of tracheids.

2. Phloem.

Phloem is a complex tissue consisting of parenchyma, sclerenchyma fibers, companion cells and sieve tube elements. Sieve tube elements are living enucleated cells through which organic materials ("food") are conducted. Sieve tube elements are arranged end to end forming a sieve tube. The protoplasts of adjacent sieve tube elements are in direct contact by means of protoplasmic strands extending through holes in their sieve plates. See Figure 23–3B. Companion cells are nucleated cells lying next to sieve tube elements.

Once more look at a slide of a *Cucurbita* stem l.s. Phloem is located just outside the xylem. Find the sieve tube elements in which the cytoplasm is stained green and an accumulation of red stained material is evident near the end walls. (Note: The red stained material may not be evident in all cells.)

Are the end walls perforated? _____

The sieve-like appearance of the end walls should be visible in some cells.

Now, examine a *Cucurbita* stem c.s. Search outside the xylem for the sieve tube cells and sieve plates. Companion cells lie next to the sieve tube cells.

V. Leaf Structure

A. External Features of Leaves.

1. Overall structure.

Leaves (see Figure 23–4) typically consist of an expanded green portion called a *blade*. The blade is usually attached to the stem by a short stalk called a *petiole*. When a petiole is absent the blade attaches directly to the stem and is termed *sessile*. The blade margin may

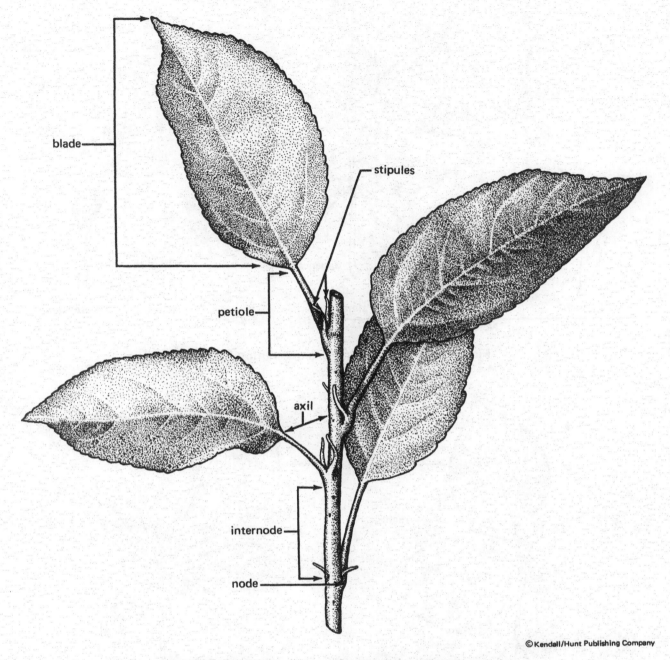

blade

stipules

petiole

axil

internode

node

Figure 23-4. A simple complete leaf showing stem relationship.

be smooth, toothed or lobed. The blade may be heart-shaped, star-shaped, linear, oval, etc. Tiny "leaflike" structures, called *stipules,* may be present at the base of the petiole. In some species of plants they are not leaflike but are modified to form spines.

2. Venation.

The arrangement of veins in a leaf is termed venation. Most monocots exhibit *parallel* venation. The main veins parallel one another. In dicots, the veins appear to intermesh forming a net. The leaf exhibits *pinnate netted venation* (the pattern is featherlike) if a single vein having lateral branches is present. The leaf exhibits *palmate netted venation* if several prominent veins arise from a common point at the base of the blade like the fingers extend from the palm of a hand. Refer to Figure 23–5A for illustrations of the basic venation patterns.

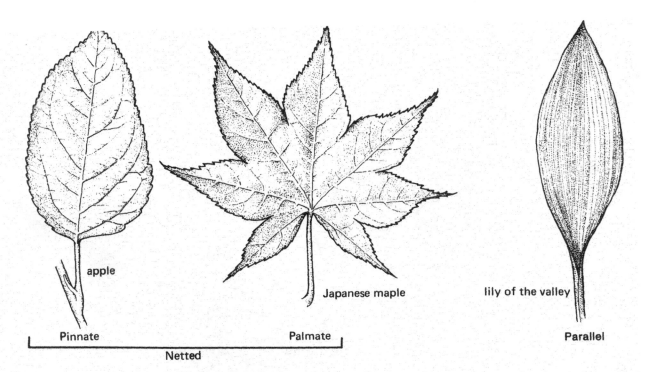

apple

Japanese maple

lily of the valley

Pinnate

Palmate

Parallel

Netted

Figure A. Simple Leaves—Basic Venation Patterns

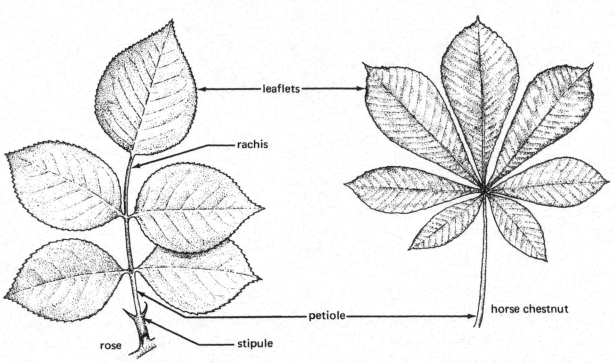

leaflets

rachis

petiole

horse chestnut

rose

stipule

Figure B. Pinnately Compound Leaf

Figure C. Palmately Compound Leaf

Figure 23-5.

3. Leaf types.

Leaves may be *simple* or *compound*. A simple leaf has a single blade. A compound leaf has three or more small blades termed *leaflets*. The distinction of simple or compound is usually made by locating the axillary bud at the base of the petiole. Leaflets (Figure 23–5B and C) may be arranged *pinnately* or *palmately*.

4. Herbarium specimens.

One or more herbarium specimens will be provided for you by your instructor. Answer the following questions about each specimen:

a. Are the leaves simple or compound? _____

b. If compound, are the leaflets pinnately or palmately arranged? _____

c. If simple, is the blade sessile or petiolate? _____

d. Is the pattern of venation parallel, pinnate netted or palmate netted? _____

e. Is the leaf margin smooth, toothed, or lobed? _____

f. What is the shape of the leaf? _____

B. Internal Features of a Dicot Leaf.

1. Basic tissues.

Although leaves vary in size and shape they all have a relatively uniform internal structure. A typical leaf (Figure 23–6) is usually flat and has three basic tissues:

a. *Epidermis*—Bricklike cells which cover the upper and lower surfaces; they usually lack chloroplasts.

b. *Mesophyll*—Consists of cells located inside the leaf in which photosynthesis occurs. What cell type makes up the mesophyll? _____

Is mesophyll a dermal, ground or vascular tissue? _____

c. *Vascular or Conducting Tissues*—Comprise the veins of a leaf which are continuous with the vascular tissue in stems and roots. They provide an avenue for transferring water and minerals from the soil to leaves and for transporting sugar from leaves to stems and roots. Vascular tissues also furnish support.

2. Preparation of leaf cross section.

Obtain a bean seedling leaf, or any typical small dicot leaf and prepare a hand section as follows:

a. Place the leaf on a clean glass slide and cut off the front one third of the leaf using a wet single edge razor blade.

b. Place your finger near the tip of the remaining portion of leaf and rest the razor blade on the leaf in front of your finger (see Figure 23–7).

c. Move your finger forward and "skip" the edge of the razor blade forward on the surface of the leaf until it almost falls off the cut edge of the leaf.

d. Slice off a thin section of leaf.

e. Float the slice in a drop of water on a slide and prepare a wet mount for observation.

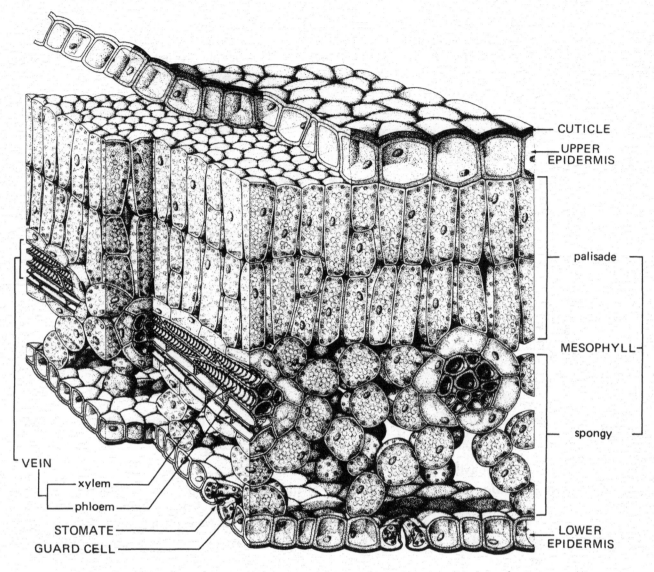

CUTICLE

UPPER EPIDERMIS

palisade

MESOPHYLL

spongy

VEIN

xylem

phloem

STOMATE

GUARD CELL

LOWER EPIDERMIS

Figure 23-6. Stereoscopic view of a dicot leaf—privet (Ligustrum).

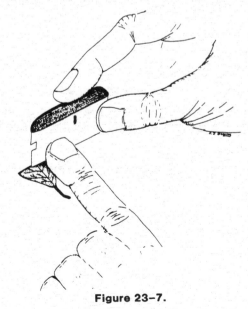

Figure 23-7.

3. Observation.

Observe your hand section and a prepared slide of a cross section of a privet (*Ligustrum*) leaf. Locate the following:

a. *Epidermis*—The epidermis is usually covered by a waxy cuticle which retards water loss from the leaf. Are there holes in the upper epidermis? _____

In the lower epidermis? _____

The holes are called stomates. On either side of a stomate is a guard cell. These are specialized epidermal cells which regulate the size of the opening. Note that the guard cells contain chloroplasts. Are chloroplasts present in other epidermal cells? _____

b. *Mesophyll*—The mesophyll can be divided into two regions:

(1) The palisade region is toward the upper surface and consists of closely packed, column-like cells.

(2) The spongy region consists of irregularly shaped, loosely packed cells. The air spaces facilitate diffusion of gases to and from the mesophyll cells of a leaf.

c. *Vein*—Notice that some veins are cut crosswise; others lengthwise. This is because of the netted venation characteristic of dicots. Each vein contains two types of vascular tissue:

(1) Xylem which conducts water and minerals and provides support, and

(2) Phloem which transports sugars.

C. Internal Features of a Monocot Leaf.

Obtain a cross section of a leaf of corn (*Zea*). Locate the following:

1. Epidermis—Are stomates present in the upper epidermis? _____ The

lower epidermis? _____

Notice the large rounded cells which are part of the epidermis. They are *bulliform cells* or "cells shaped like bubbles." At least three possible functions have been suggested for them:

a. Storage of water.

b. Changes in turgor pressure in them plays a role in the opening and closing movements of mature leaves. Rolling up of the leaves favors a reduction in the amount of transpiration.

c. Unfolding of the blade during early development results from their expansion.

2. Mesophyll—Notice that the mesophyll is not differentiated into palisade and spongy regions as in dicots.

3. Veins—Observe that all of the veins are cut crosswise. This is a reflection of the parallel venation characteristic of monocots. Also, note the bundle sheath cells surrounding the smaller

veins. Were bundle sheath cells present in the dicot leaf? _____ .

D. Stomates

Stomates are holes in leaves through which carbon dioxide can enter the leaf from the surrounding atmosphere and water and oxygen can escape from the leaf. For a face view of a stomate:

271

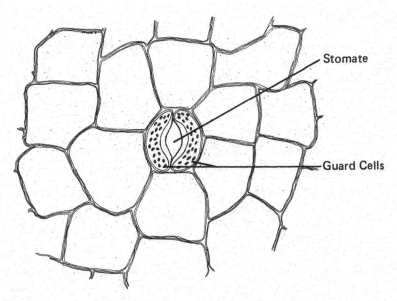

Figure 23–8. Stomatal apparatus, face view.

1. Obtain a lettuce (*Lactuca*) leaf, and tear off part of the epidermis. Do this by snapping the leaf along a vein and peeling back the epidermis. The epidermis should look like a piece of cellophane.

2. Prepare a wet mount of a small piece of epidermis.

3. Locate the bean shaped guard cells and the stomate (see Figure 23–8). Remember that the only epidermal cells that contain chloroplasts are guard cells.

To obtain variation in guard cells, obtain part of a corn seedling leaf. Place the leaf on a clean glass slide. Using a razor blade, scrape a thin area in one section of the leaf, cover it with water and prepare a wet mount. Locate and sketch the stomate and guard cells.

VI. Photosynthesis

In photosynthesis, green plants manufacture carbohydrates and oxygen in cell organelles called chloroplasts utilizing simple inorganic raw materials (H_2O and CO_2), radiant energy from the sun and chloroplast enzymes and pigments. This process is summarized by the following unbalanced equation:

$$CO_2 + H_2O \xrightarrow[\text{Chloroplast}]{\text{Light}} C_6H_{12}O_6 + O_2$$

Carbon dioxide   Water   Glucose   Oxygen

In higher plants, most of the chloroplasts are located in leaves. Therefore, most of the photosynthesis occurs in them.

A. Chloroplast Pigments Absorb Light Energy

Chloroplast pigments absorb the light that makes the process of photosynthesis possible. The main pigment in chloroplasts is chlorophyll a. Other pigments, including chlorophyll b, the carotenes, and the xanthophylls play secondary roles absorbing light of different wave lengths than chlorophyll a and transferring the energy they absorb to chlorophyll a for direct use in the light reactions of photosynthesis. These pigments can be extracted from green plant tissues with lipid solvents, and separated on chromatography paper. The separation of the mixture of pigments is based on differences in the solubilities of the individual pigments in different solvents, differences in the affinity of the individual pigments for the chromatography paper, and differences in the molecular weight of the pigments.

B. Paper Chromatography of Chloroplast Pigments

1. Using a mortar and pestle, prepare a pigment extract by grinding two or three spinach leaves (or other suitable leaves) with sand and acetone. Use as little acetone as possible so that the pigment extract is as concentrated as possible.

2. With a capillary tube apply a narrow strip of the extract near the pointed end of the paper (see Figure 23–9).

3. Apply the extract on the same line five or six more times, drying the paper after each application by waving it in the air.

4. Suspend the chromatography paper in a test tube containing enough petroleum ether (about one-quarter inch) to keep the tip of it wet. Be careful that the paper does not touch the edge of the test tube and that the petroleum ether is not splashed up on it.

5. Allow the chromatogram to develop until the pigments are separated. When the solvent front reaches the top of the paper, remove the strip, let it dry and examine it. The yellow band of *carotenes* moves fastest. The yellow-brown *xanthophyll* band will be lower, followed by the bluish-green *chlorophyll a* and the lowest yellow-green band of *chlorophyll b*. Mark and label location of bands on Fig. 23–9.

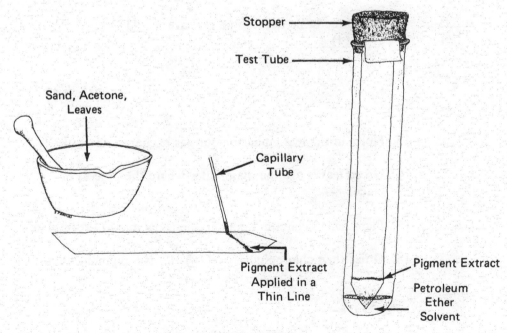

Figure 23–9.

273

_____ 1. A region of cell division is termed a(n) ___(1)___ .

_____ 2. The apical meristem of a root is protected by a(n) ___(2)___ .

_____ 3. The layer that forms the outer covering of leaves and young roots and stems is ___(3)___ .

_____ 4. Two types of conducting cells may be present in xylem. They are ___(4)___ and ___(5)___ .

_____ 5.

_____ 6. A Casparian strip is characteristic of ___(6)___ .

_____ 7. ___(7)___ consists of living cells with unevenly thickened primary walls and functions in support.

_____ 8. Because of their thick walls the conducting cells of xylem also function in providing ___(8)___ .

_____ 9. The conducting cells of phloem are called ___(9)___ .

_____ 10. Areas in the wall of a fiber with little or no secondary thickening are termed ___(10)___ .

_____ 11. A simple leaf can be distinguished from a compound leaf by locating the position of the ___(11)___ .

_____ 12. The three basic tissues in a leaf are ___(12)___ , ___(13)___ and ___(14)___ .

_____ 13.

_____ 14.

_____ 15. Holes in the leaf epidermis for gas exchange are called ___(15)___ .

_____ 16. Green leaves are the major sites where the process of ___(16)___ takes place.

_____ 17. Plant leaves contain ___(17)___ (how many) green pigments.

_____ 18. Other than chlorophyll, ___(18)___ is a pigment found in green leaves.

STEM AND ROOT ANATOMY

Objectives

Upon completion of this exercise you should be able to:

1. Distinguish between primary and secondary (lateral) meristems.
2. Describe the differences between stem and root tips as seen microscopically.
3. Locate and identify microscopically the primary tissues of a:
   a. herbaceous dicot stem
   b. herbaceous dicot root.
4. Locate and identify microscopically the secondary tissues of a woody dicot stem.
5. Relate the terms bark and periderm to the secondary tissues of a woody dicot stem.
6. Name the lateral meristems from which each secondary tissue of a woody dicot stem develops.
7. Determine how old a woody stem was when it was cut down by examining the cross section.
8. Recognize the following features on a woody twig:
   terminal bud
   axillary bud
   leaf scar
   vascular bundle scars
   bud scale scars
   lenticels
   lateral branches
   leaf arrangement.
9. Determine the age of a young lateral branch.
10. Recognize a branch root on a slide of a root cross section and name the primary tissue that gives rise to it.
11. Answer all questions in the exercise.

I. Stem Structure

A. The Dicot Stem Tip (Figure 24–1).

At the tip of the stem is a region of actively dividing cells, a meristem, which produces cells in a direction away from the tip of the stem. The *apical meristem* is responsible for growth in length of the stem. The tissues formed are *primary tissues*.

Two types of lateral appendages, *leaves* and *buds*, are produced in the area of mitotic divisions in the stem apical meristem. The primordial (developing) leaves and buds undergo their own sequence of development aside from the main axis development of primary tissue.

Obtain a slide of *Coleus* stem tip l.s. Locate leaf and bud primordia and the apical meristem. Note the position of the buds. They are situated in the angle between the stem and the leaf. This is the *axil* of the leaf. The buds are termed *axillary buds*.

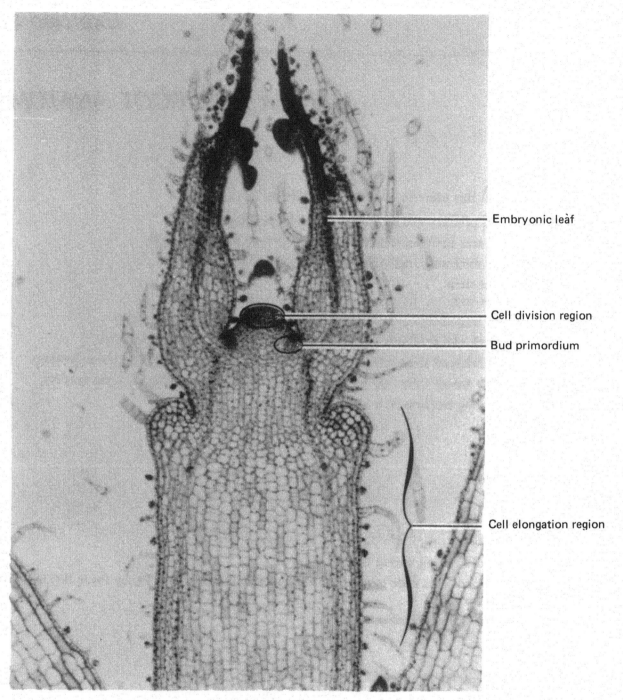

Embryonic leaf

Cell division region

Bud primordium

Cell elongation region

Figure 24-1. *Coleus* stem tip, l.s.

At some distance behind the stem tip, cells derived from the apical meristem begin to elongate (enlarge) and differentiate to form primary tissues. Can you see any evidence of cell elongation and/or differentiation behind the apical meristem? _____

_____

_____

Is the white "Irish" potato a stem? _____

The answer is yes. The "eyes" are axillary buds and even though a potato forms below ground, it is a stem.

B. Primary Tissues of a Dicot Stem.

The mature tissues formed by elongation and differentiation of cells derived from the apical meristem are primary tissues. Primary tissues form the primary body.

1. Preparation of a stem cross section.

Obtain a bean seedling and place the root end in a beaker containing eosin dye for 15 minutes. Remove the seedling and wash it free of dye. Prepare a stem cross section as follows:

a. First cut a short section of stem using a razor blade.

b. Put the piece of stem on a clean glass slide.

c. Slice a "thin" cross section from either cut end.

d. Prepare a wet mount and examine the section.

e. Repeat the procedure until you obtain a thin section suitable for examining. Check with your instructor if you are uncertain.

Where is the red dye concentrated? _____

The red dye is transported upward in the primary xylem. Some lateral conduction into other primary tissues may occur. Show the distribution of primary xylem in the diagram below.

2. Identification of primary tissues.

Locate each primary tissue described below on the section you prepared.

a. *Primary xylem.*

The primary xylem in a dicot stem occurs in separate vascular bundles arranged in a circle.

b. *Primary phloem.*

Primary phloem is located in the same vascular bundles outside the primary xylem. Between the primary phloem and primary xylem is a region of undifferentiated cells. These give rise to the *vascular cambium* of woody plants.

c. *Sclerenchyma fibers.*

Collectively fibers may form a "cap" over each vascular bundle. The cell wall is thick and the cell lumen is narrow.

d. *Pith.*

Pith is a region of cells located in the center of the stem axis. Pith is composed of parenchyma cells that function in starch storage.

e. *Pith rays.*

Pith rays consist of parenchyma cells located between the vascular bundles.

f. *Cortex.*

Cortex is the region, several cell layers thick, located beneath the epidermis and to the outside of the vascular bundles. It consists mainly of living parenchyma cells which usually store starch.

Is the stem green? _____ If so, in what cells are the chloroplasts located?

_____

g. *Epidermis.*

Epidermis forms the outer covering of the primary body of the stem. It is usually one cell layer thick, consists of living cells, and may have a waxy coating called the *cuticle,* which retards water loss through the epidermis. Hairs may also be present on the epidermis. Epidermal cells generally lack chloroplasts. Guard cells are an exception.

After locating each of the primary tissues in the hand section you made, obtain and observe a prepared slide of a stem cross section of alfalfa (*Medicago*) or of a sunflower (*Helianthus*). Locate each of the primary tissues mentioned above. Use Figures 24–2 and 24–3 to verify your observations.

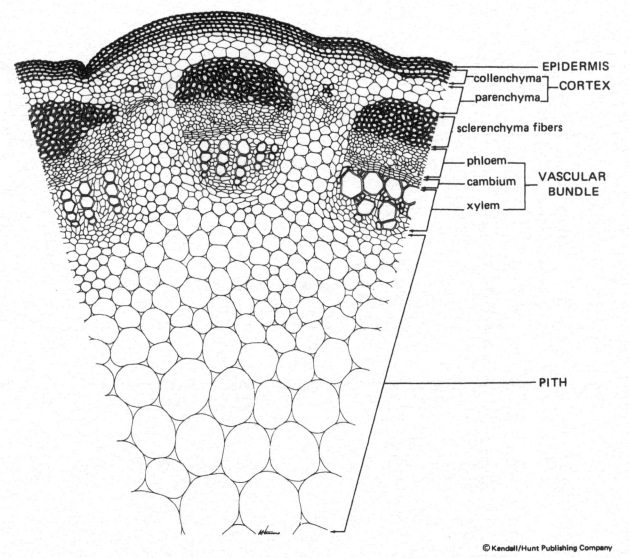

Figure 24–2. The herbaceous dicot stem cross section of Helianthus (sunflower) showing primary tissues.

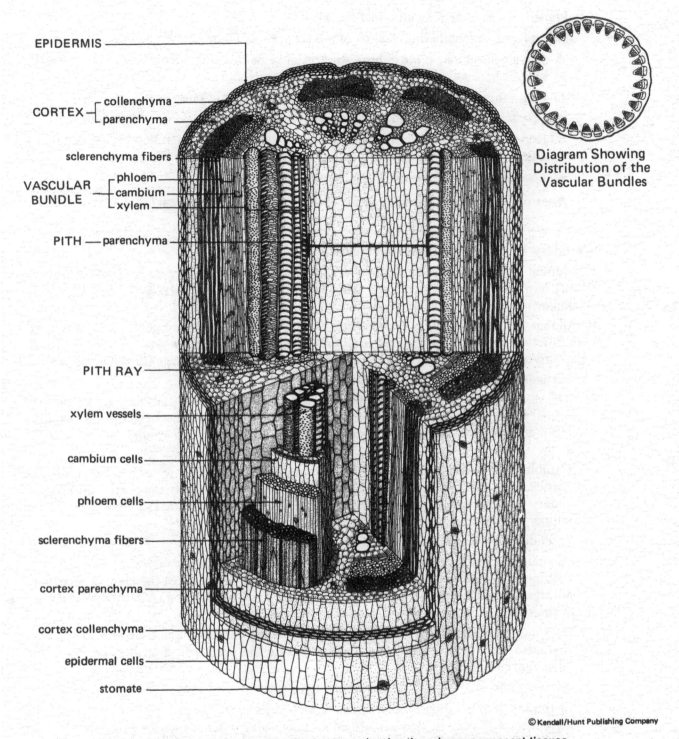

EPIDERMIS

CORTEX — collenchyma
         parenchyma

sclerenchyma fibers

VASCULAR — phloem
BUNDLE   cambium
           xylem

PITH — parenchyma

PITH RAY

xylem vessels

cambium cells

phloem cells

sclerenchyma fibers

cortex parenchyma

cortex collenchyma

epidermal cells

stomate

Diagram Showing
Distribution of the
Vascular Bundles

© Kendall/Hunt Publishing Company

Figure 24–3. The herbaceous dicot stem—showing the primary permanent tissues.

3. Preparation of a stem longitudinal section.

Now prepare a longitudinal section of a bean stem as follows:

a. Cut a one cm long piece of bean stem on an angle to divide it lengthwise into two wedge shaped halves.

b. Cut a thin longitudinal section from the angular face of one piece.

c. Prepare a wet mount and examine the section.

d. Repeat the procedure until you obtain a thin longitudinal section suitable for examining. Check with your instructor if you are uncertain.

e. Locate each of the primary tissues mentioned above.

What patterns of wall thickenings are evident in the primary xylem? _____

_____

C. Secondary Tissues of a Dicot Stem (Figure 24–4).

In addition to the primary body, many plants also form a secondary body. These plants are usually termed woody. Both roots and stems can have a secondary body. Because of their overall similarity, only the secondary body of a stem will be considered.

Recall that the primary body is derived from the apical meristem. The secondary body originates from mitotic divisions of two lateral meristems, the *vascular cambium* and the *cork cambium*. Cells derived from lateral meristems produce an increase in the diameter (thickness) of a stem.

1. Secondary growth from the vascular cambium.

The vascular cambium is derived from relatively undifferentiated cells located between the primary xylem and phloem and from pith ray parenchyma cells in between vascular bundles. Ultimately, cells of the vascular cambium form a continuous ring around the inside of the stem.

Cambial cells divide forming secondary xylem to the inside and secondary phloem to the outside. New cambial cells and vascular rays are also produced by the cambium. The secondary xylem produced is located outside the primary xylem. Primary phloem may persist temporarily in small islands outside the secondary phloem.

2. Secondary growth from the cork cambium.

Annual increments of secondary xylem force all other tissues outward. As a consequence, the ring of cambium continually increases in circumference as does the layer of secondary phloem. Also, the epidermis, cortex and primary phloem slough off. A second lateral meristem, cork cambium, differentiates within the primary phloem.

Cork cambial cells (Figure 24–5B) divide producing cork to the outside and cork parenchyma to the inside. Collectively, they constitute the *periderm*. Together, periderm and phloem are termed bark.

3. Annual rings.

In the temperate region an increment of secondary xylem is laid down each year. Each increment of secondary xylem is termed an annual ring. Each annual ring consists of spring and summer wood. Because the conducting cells of spring wood are larger in diameter than those of summer wood, one annual ring can be distinguished from another. By counting the number of annual rings, the approximate age of a tree can be determined.

Obtain cross sections of one and three year old basswood (*Tilia*) stems. From the center proceeding toward the outside identify pith, primary xylem, secondary xylem, vascular cambium, secondary phloem, vascular ray, and periderm. Also, distinguish between the spring and summer wood of one annual ring.

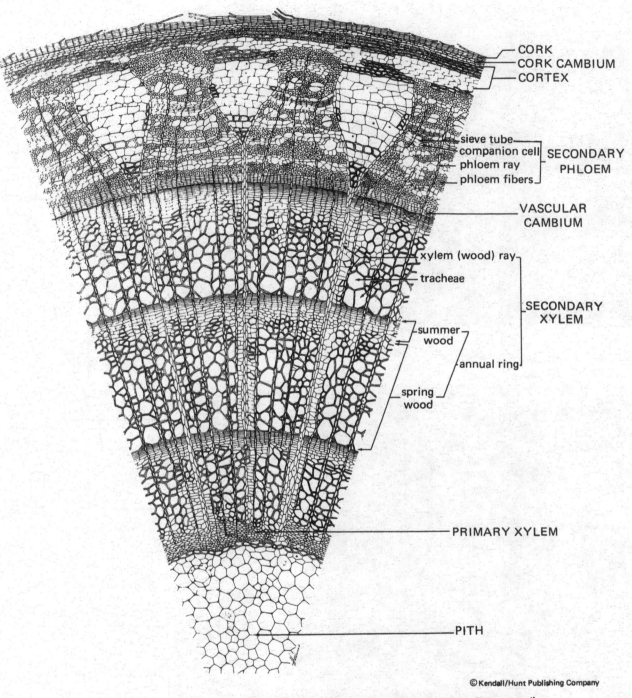

CORK
CORK CAMBIUM
CORTEX

sieve tube
companion cell
phloem ray
phloem fibers
} SECONDARY PHLOEM

VASCULAR CAMBIUM

xylem (wood) ray
tracheae

SECONDARY XYLEM

summer wood
annual ring
spring wood

PRIMARY XYLEM

PITH

© Kendall/Hunt Publishing Company

Figure 24-4. Cross section of a woody dicot stem 3-year old Tilia (basswood).

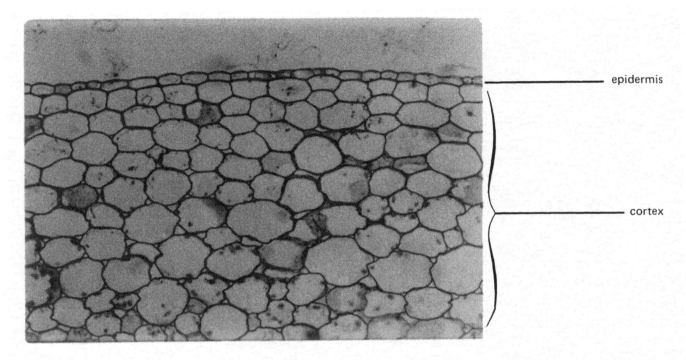

epidermis

cortex

A. Young *Geranium* Stem x.s.

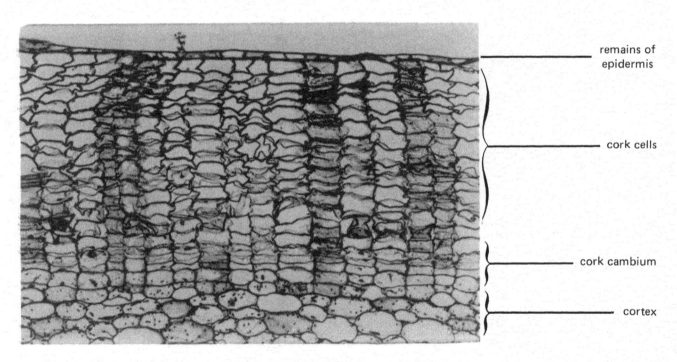

remains of epidermis

cork cells

cork cambium

cortex

B. Older *Geranium* Stem x.s.

Figure 24–5.

282

Next, determine the age of a piece of wood. _____ Check the answer with your instructor.

Now, obtain cross sections of young and old stems of *Geranium*. Identify the epidermis and cortex of a young stem. Then, identify the cork cells and cork cambium of an older stem. Check your observations against Figures 24–5A and 24–5B.

D. The Monocot Stem.

Monocot stems differ from dicot stems in two ways:

1. A cambium is absent.

2. The vascular bundles are scattered throughout the stem instead of being arranged in a circle.

Obtain a cross section of corn (*Zea mays*). Show the arrangement of vascular bundles in the diagram below.

E. External Features of Stems (Figure 24–6).

Several structures may be found on both herbaceous and young woody stems. The point of leaf attachment to the stem is called a *node*. Leaf arrangement is determined by the number of leaves produced at a node. One leaf present at a node is the *alternate* arrangement; two per node are the *opposite* arrangement, and three or more per node are the *whorled* arrangement.

The portion of a stem between two consecutive nodes is called an *internode*. Another feature is the potential growing points called *buds*. Each bud contains a meristem which is usually covered by thick modified leaves called *bud scales*. Buds may be at the tip of a stem (*terminal*) or in the axil of a leaf just above the node, (*axillary* or *lateral*).

Several features are found only on woody stems. These structures are due to the very nature of a woody plant in that they represent remnants of structures no longer present or structures newly acquired because of the perennial nature of woody plants. *Bud scale scars* are left on the twig when bud scales fall off. Within limits, the age of a lateral branch can be determined by locating the bud scale scars. The portion of the stem between two successive areas of bud scale scars represents one year's growth.

The *vascular bundle scar* is found within the *leaf scar* and is a remnant of the vascular tissue which connected the stem to the leaf.

Herbaceous stems and young woody stems are covered by a single layer of cells, the epidermis. In older woody stems a many-celled layer called *bark* covers the stem. In order for oxygen to get into the living cells of the woody stem and for carbon dioxide to get out, there are openings in the bark called *lenticels*. Lenticels are usually visible to the unaided eye. They appear as tiny specks on the bark.

Obtain a woody twig. On it locate terminal and axillary buds, leaf scars, vascular bundle scars, bud scale scars, lenticels and lateral branches. How many years of growth are represented on

the main axis of the twig? _____

What structures give rise to lateral branches? _____

Was the leaf arrangement alternate, opposite or whorled? _____

Check the above answers with your instructor.

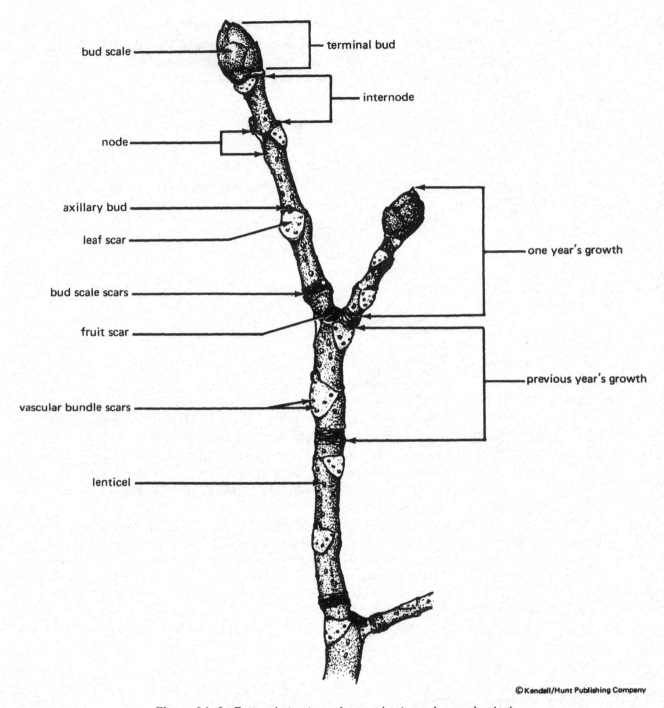

bud scale

terminal bud

internode

node

axillary bud

leaf scar

bud scale scars

fruit scar

one year's growth

previous year's growth

vascular bundle scars

lenticel

Figure 24-6. External structure of a woody stem—horse chestnut.

F. Specialized Stems (Figure 24–7).

    1. *Stem succulents*—In many plants the stem is large and fleshy. It stores large quantities of water, is green and is able to manufacture food. Succulents get through dry periods by utilizing the stored water. Since many succulents are susceptible to freezing, they commonly occur in subtropical deserts. Examples include cacti and euphorbs.

    2. *Rhizomes* are horizontal, perennial underground stems. Leaves are produced at the nodes on aerial branches arising from buds. Roots may also be produced from the nodes. Examples are found in cattails, orchids and irises.

    3. *Tubers* are enlarged, stocky, fleshy underground stems, usually found on a rhizome. The white potato is an example.

    4. *Bulbs* are short, vertical underground stems which have fleshy leaves. Examples are found in lilies and onions.

    5. *Corms* are short, flat, vertical, much thickened underground stems. Examples are found in crocus and gladiolus.

II. Root Structure

A. Dicot Root Tip (Figure 24–8).

Recall having examined a longitudinal section of an onion (*Allium*) root tip. How did it differ from the longitudinal section of a *Coleus* stem tip?

    1. _____

    2. _____

The root apical meristem produces new cells, primarily away from the tip, which enlarge (elongate) and differentiate to form primary tissues. The newly added cells are responsible for the growth in length of the root. Covering the apical meristem of the root is a collection of nondividing cells called the *root cap*. Root cap cells also arise from the apical meristem. No lateral appendages (e.g., leaves and buds) are produced at the root tip.

B. Primary Tissues of a Dicot Root (Figure 24–9).

The primary body of a dicot root consists of primary tissues derived from the apical meristem. Obtain a slide of a buttercup (*Ranunculus*) root cross section. Locate the following tissues:

    1. *Epidermis.*

    Epidermis forms the outer covering of a root. It is usually one cell layer thick and consists of living cells. Epidermal cells may have lateral projections which increase the absorptive surface area of the root. These projections are called root hairs.

    2. *Cortex.*

    Cortex is located inside the epidermis. It is a region, several cell layers thick, consisting of parenchyma cells that usually store starch.

    3. *Endodermis.*

    Endodermis is the innermost layer of the cortex. Can you see Casparian strips?

    _____What is their function? _____

    4. *Pericycle.*

    The pericycle is a layer of living cells inside the endodermis. By mitotic cell division, elongation and differentiation it gives rise to lateral or branch roots, cork cambium and some vascular cambium. The remaining vascular cambium, formed in plants having secondary growth, differentiates from cells located between the primary phloem and primary xylem.

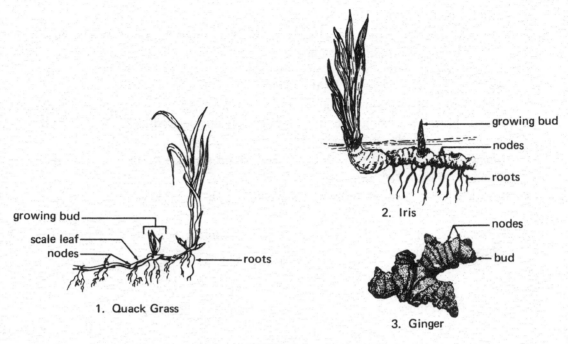

1. Quack Grass

growing bud
scale leaf
nodes
roots

2. Iris

growing bud
nodes
roots

nodes
bud

3. Ginger

A. Rhizomes

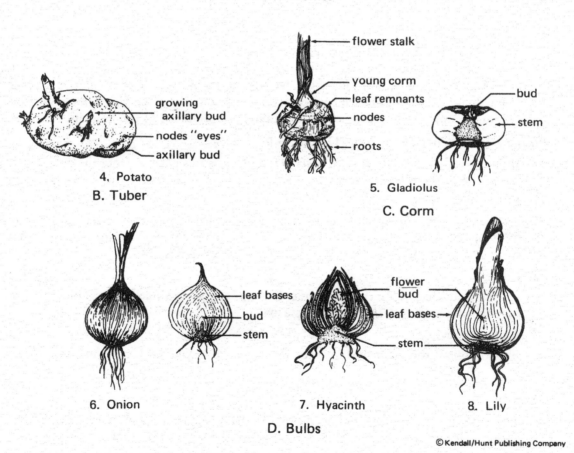

growing
axillary bud
nodes "eyes"
axillary bud

4. Potato
B. Tuber

flower stalk
young corm
leaf remnants
nodes
roots

bud
stem

5. Gladiolus
C. Corm

leaf bases
bud
stem

flower
bud
leaf bases
stem

6. Onion

7. Hyacinth

8. Lily

D. Bulbs

Figure 24-7. Types of specialized stems—underground.

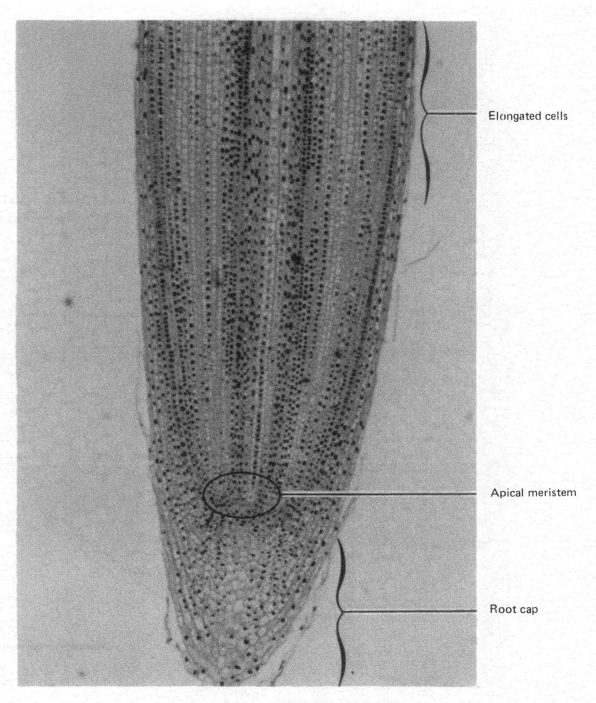

Elongated cells

Apical meristem

Root cap

Figure 24-8. *Allium* (onion) root tip, l.s.

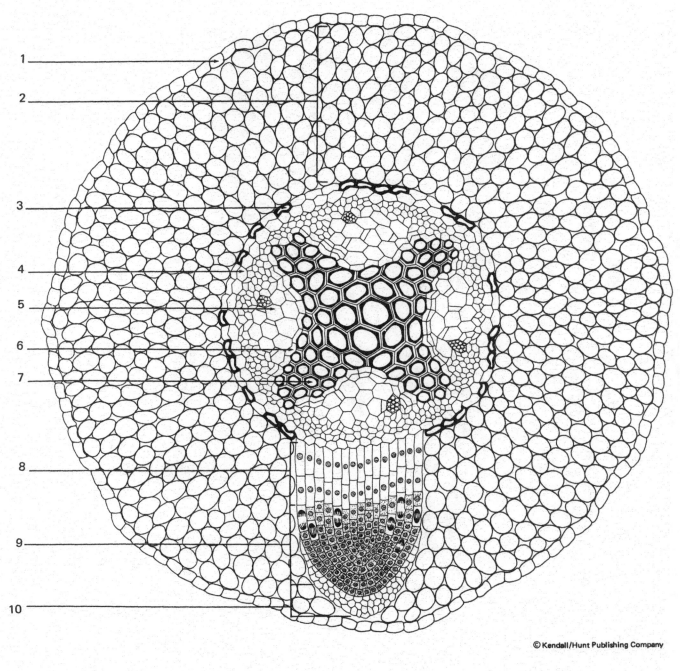

1. Epidermis
2. Cortex
3. Endodermis
4. Pericycle
5. Phloem
6. Cambium
7. Xylem
8. Region of Cell Elongation
9. Region of Cell Division
10. Root Cap

Figure 24-9. Cross section of a root tip—origin of a secondary root.

5. *Primary phloem.*

   Primary phloem is found inside the pericycle in several small "islands" (usually 3 or 4) located between extensions of primary xylem. Phloem consists of sieve tube elements, companion cells, parenchyma cells and fibers. Sugar conduction occurs in phloem.

6. *Primary xylem.*

   Primary xylem occupies the center of the root cross section. Several arms (2-4, usually 4 in the shape of an X) of it extend between "islands" of primary phloem. Xylem consists of vessel elements, parenchyma cells and fibers. Water and minerals move upward in the vessels.

C. The Absorptive Surface of the Root.

   Water and minerals present in the soil environment enter the plant at the root. Two features of roots enable them to expose to the soil environment an extensive surface area through which the absorption of water and minerals can occur. These are root hairs and lateral or branch roots. Recall that root hairs are lateral extensions of epidermal cells. Would you expect to find root hairs

   on older portions of a root? _____

   What happens to the stem epidermis during secondary growth? _____

   _____

   If you don't remember, refer back to the section on the secondary tissues of a dicot stem.

   Obtain a germinated radish *(Raphanus)* seed. Look at the root hairs under a stereo-microscope. Are root hairs present at the extreme tip where cells are dividing? _____

   Where are they located? _____

   Observe a longitudinal section of corn *(Zea)* root or refer to Figure 24-10 if you cannot answer this question.

   Lateral or branch roots are formed by division, elongation and differentiation of cells of the pericycle. Look at the slide of a willow *(Salix)* root cross section. What tissues are pushed out

   of the way when a lateral root originates from the pericycle? _____,

   _____ and _____. For aid in answering the question refer to Figure 24-9.

D. The Monocot Root.

   Pith is present in the center of the root in many monocots and some herbaceous dicots. The number of arms of primary xylem present in a monocot root cross section usually exceeds the number present in dicot roots and may reach 20.

   Obtain a cross section of a corn *(Zea)* root. Does primary xylem or pith occupy the center of the

   corn root? _____ How many arms of primary xylem are present?

   _____

   On the same slide identify epidermis, cortex, endodermis, pericycle and primary phloem.

E. Types of Roots (Figures 24-11 and 24-12).

   The root which arises from the lower end of the embryo during seed germination is called the primary root. Lateral or branch roots are termed secondary if they arise from the primary root, tertiary if they arise from a secondary root and so on.

1. *Taproot system*—If the primary root persists as the main root it is called a taproot. Beets, carrots, dandelions and sunflowers have taproots.

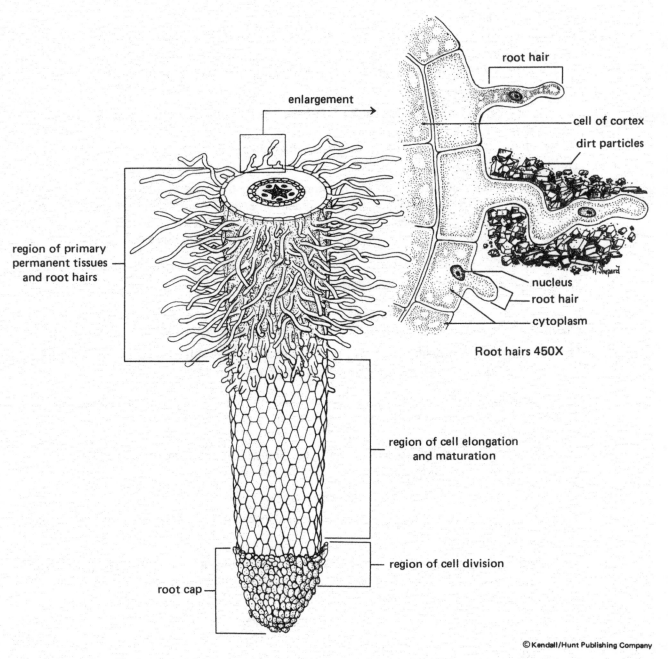

enlargement

root hair

cell of cortex

dirt particles

nucleus

root hair

cytoplasm

Root hairs 450X

region of primary
permanent tissues
and root hairs

region of cell elongation
and maturation

region of cell division

root cap

© Kendall/Hunt Publishing Company

Figure 24-10. Structure of a primary root—external showing root hair structure.

2. *Fibrous root system*—In a fibrous root system, many roots arise from near the base of the stem and all of them are about the same size. Dahlia, grasses, sweet potatoes and wheat have fibrous root systems.

3. *Adventitious roots*—Roots that grow from stems or leaves are termed adventitious. The *aerial roots* of English ivy and poison ivy are examples as are the *prop roots* of corn and mangroves. Also, roots that arise from stem and leaf cuttings are adventitious.

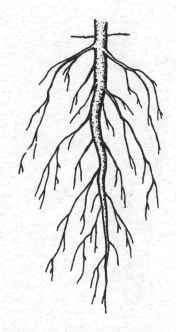

A. Fibrous Root System of Wheat

B. Taproot System of Sunflower

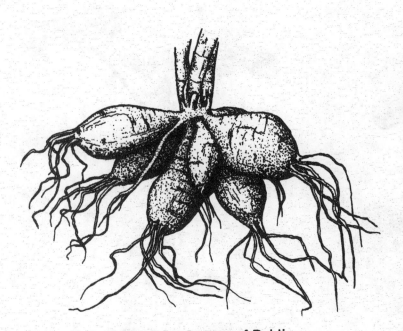

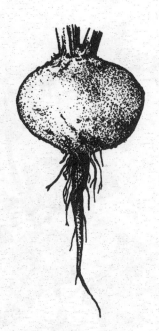

C. Fleshy Root System of Dahlia

D. Fleshy Taproot System of Beet

Figure 24–11. Types of root systems.

291

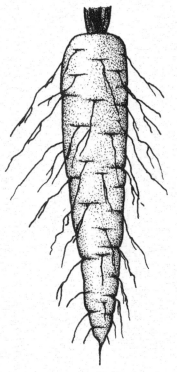

A. Food Storage Root of Carrot

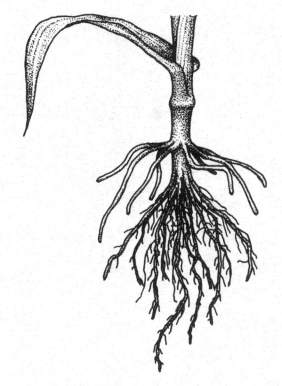

B. Prop Roots of Corn

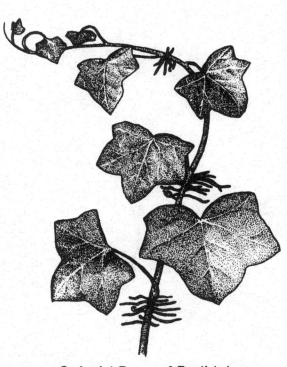

C. Aerial Roots of English Ivy

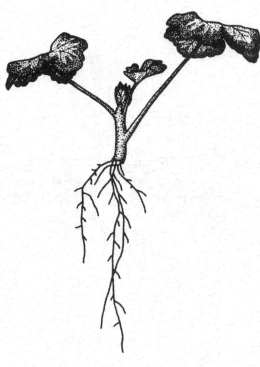

D. Adventitious Roots on Cutting of
Cultivated Geranium

Figure 24-12. Specialized roots.

REVIEW QUESTIONS

_____ 1. ____(1)____ and ____(2)____ primordia arise in the stem apical meristem.

_____ 2.

_____ 3. The conducting tissues in a herbaceous dicot stem are located in separate ____(3)____ arranged in a circle.

_____ 4. The primary tissue in the center of a herbaceous dicot stem is ____(4)____ .

_____ 5. The secondary tissues of a woody stem originate in two lateral meristems. They are the ____(5)____ and ____(6)____ cambia.

_____ 6.

_____ 7. Bark consists of periderm and ____(7)____ .

_____ 8. Openings in the bark for gas exchange are called ____(8)____

_____ 9. Root hairs are lateral extensions of ____(9)____ cells.

_____ 10. Lateral or branch roots originate from the ____(10)____ .

11. Match each stem tissue or organ listed on the left with the meristem from which it originates.

_____ Periderm   A. apical meristem

_____ Pith   B. cork cambium

_____ Primary phloem   C. vascular cambium

_____ Leaf

_____ Epidermis

_____ Secondary xylem

12. Match each root tissue or structure listed on the left with the meristematic area from which it originates.

_____ Root cap   A. apical meristem

_____ Root hair   B. cork cambium

_____ Endodermis   C. pericycle

_____ Branch or lateral root   D. vascular cambium

_____ Periderm

REPRODUCTION AND DEVELOPMENT
OF FLOWERING PLANTS

Objectives

Upon completion of this exercise you should be able to:

1. Identify and list the function of the four main parts of a typical flower.
2. Determine the location of egg, ovule and ovary in a flower.
3. Define or recognize where applicable, the following processes or structures in relation to reproduction in the flowering plants:

pistil (stigma, style and ovary)	fruit
ovule	sporophyte
embryo sac	heterospory
egg	stamen (anther and filament)
polar nuclei	pollen
double fertilization	pollination
seed	

4. Describe the relationship between each of the following: integument and seed coat, seed and ovule, fruit and ovary.
5. Identify and locate on a lima bean seed—hilum, micropyle, seed coat, cotyledon, epicotyl, hypocotyl and radicle.
6. Distinguish between simple, aggregate and multiple fruits.
7. State two functions for seeds and fruits.
8. Contrast seedling development in lima beans, peas and corn.
9. List at least three features which have enabled flowering plants to become the dominant forms of plant life on land.
10. Distinguish between tropisms and nastic movements.
11. Identify the action of specified plant hormones.
12. Answer all questions in the exercise.

I. Introduction

Flowering plants can live one, two or an indefinite number of years. To survive, those that live for an indefinite number of years must remain free of disease and accident, and must maintain an uninterrupted supply of water, minerals, light, etc. Eventually, even they will die. When individual plants of a particular kind (i.e., maples, oaks, willows, etc.) die they are usually replaced by other individuals of the same type. Generally, the new individuals begin their existence inside flowers following sexual reproduction. The basic life cycle of a flowering plant is shown in Figures 25–1 and 25–3.

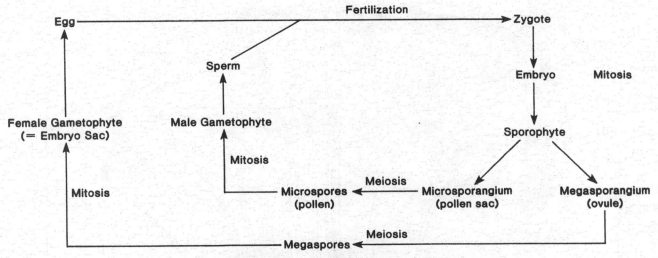

Figure 25-1. Basic life cycle of a flowering plant.

II. Angiosperms (Flowering Plants)

The sporophyte is the conspicuous part of the life cycle and it bears the flowers and fruits which are distinctive features of this group. Flowering plants are heterosporous and, therefore, produce two types of gametophytes. Pollination is accomplished by wind, insects, water, bats, birds, etc. Nonmotile sperm are produced.

A. *Angiosperm Sporophyte*—

The mature sporophyte is differentiated into roots, stems and leaves. Roots are involved in the absorption of water and minerals and are specialized for anchoring the plant body to the soil which is an advantage in a terrestrial environment. Stems support the leaves and provide an avenue for conduction of materials to and from the leaves. The leaves are specialized for carrying on photosynthesis.

1. The Flower (Figure 25–2)

   The flower is an organ adapted for sexual reproduction. Some flowers are valued highly by humans for their aesthetic value. They vary considerably in beauty, color, form, size and number of parts. This diversity ranges from the showy orchids and roses to the relatively inconspicuous flowers of grasses and oaks. Typically, a flower consists of four types of flower parts attached in whorls or concentric rings to the expanded end, *receptacle,* of a flower stalk. Flower parts can be divided into those directly involved in the reproductive process (essential) including stamens and pistils, and those not directly involved (nonessential) including sepals and petals. Identify each part on a flower model and then on fresh or preserved flowers.

   a. *Sepals*—these modified leaves may be green or some other color. They may be separate or fused into a tube.

   b. *Petals*—petals are usually colored. Colored sepals and petals serve to attract insects. Petals, like sepals, may be separate or fused.

   c. *Stamens*—a flower usually has two or more stamens. Each stamen consists of a *filament* or stalk that bears an *anther* at its tip. The anther contains chambers in which *pollen* is produced. Male gametes or sperm are produced in pollen.

   d. *Pistils*—each pistil consists of one or more modified leaves or *carpels*. A *simple pistil* consists of one carpel. A *compound pistil* consists of two or more fused carpels.

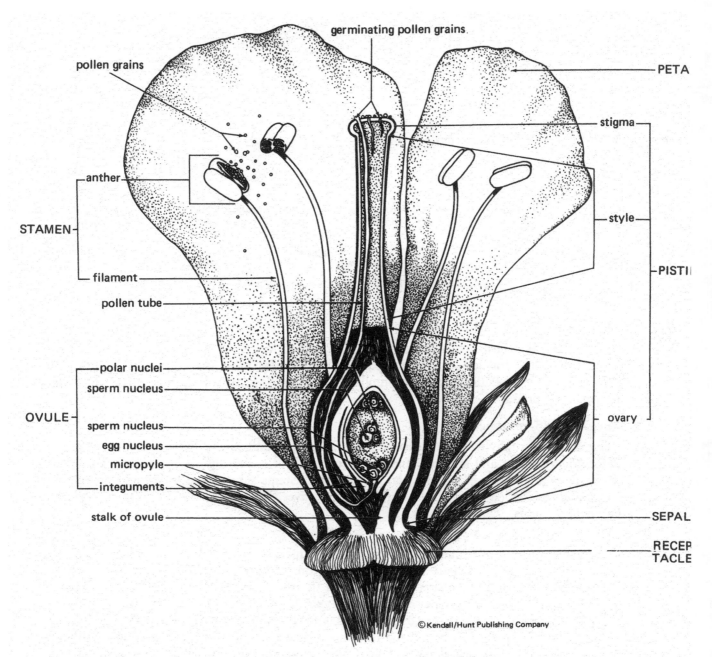

pollen grains

germinating pollen grains

PETA

stigma

anther

style

STAMEN

PISTI

filament

pollen tube

polar nuclei
sperm nucleus

OVULE

sperm nucleus
egg nucleus
micropyle
integuments

ovary

stalk of ovule

SEPAL

RECEP
TACLE

© Kendall/Hunt Publishing Company

Figure 25-2. Generalized structure of a dicot flower.

The carpel of a simple pistil consists of stigma, style and ovary. The *stigma* is the upper part of the pistil that is usually sticky and provides a receptive surface for the pollen. The *style* connects stigma to ovary. The ovary is the enlarged lower portion of the pistil. It usually contains one or more *ovules*. Each ovule is surrounded by two protective layers or *integuments* except for a small opening at one end called the *micropyle*. An egg is located inside each ovule near the micropyle.

B. *Angiosperm Gametophytes*

Flowering plants are heterosporous. They form two types of spores that subsequently form two types of gametophytes.

1. *Angiosperm Male Gametophyte*—the male gametophyte has reached the ultimate in reduction of the gametophyte. It consists of two cells inside the pollen. One cell, the generative cell divides prior to fertilization to form two nonmotile sperm.

   Examine a slide labeled *Lilium* pollen and locate the two nuclei (generative cell and tube cell), which make up the male gametophyte. Trace the development in Figure 25–3, numbers 13–21.

2. *Angiosperm Female Gametophyte*—the female gametophyte of the lily consists of eight nuclei inside an oval or elongated embryo sac located in an ovule (sporangium). Trace the development of the embryo sac using models if available or use Figure 25–3, numbers 22–42. Within the ovule a diploid cell undergoes meiosis to produce four haploid cells. Three of the haploid cells disintegrate. The remaining haploid cell gives rise to the embryo sac in which the embryo develops. The haploid nucleus divides three times by mitosis to form eight genetically identical nuclei. How are the eight nuclei arranged in the embryo sac?

   One of the haploid nuclei nearest the opening into the ovule (the micropyle) serves as the egg. Two of the nuclei (polar nuclei) move to the center of the embryo sac. Examine a slide labeled *Lilium* ovary: mature embryo sac. First, using the scanning power, locate the three pairs of ovules. Make a simple sketch showing the relative positions of these ovules.

Select an ovule with a well defined embryo sac and observe it under low magnification. Note the two layers of cells which appear to wrap around the embryo sac. What are they called?

_____ Now switch to high power and locate as many of the eight nuclei of the embryo sac as you can. Note that it is very rare that one microscopic cross section contains all eight nuclei. Make a sketch of the embryo sac and the two surrounding integument layers. Label the egg and the two polar nuclei.

Following pollination, a pollen tube grows through the style and ovary toward an ovule. The generative cell divides to form two nonmotile sperm cells.

One of the sperm cells unites with the egg to form a zygote. The other sperm cell unites with the two polar nuclei resulting in the production of triploid endosperm. This process, called double fertilization, is unique to angiosperms. The endosperm nourishes the embryo.

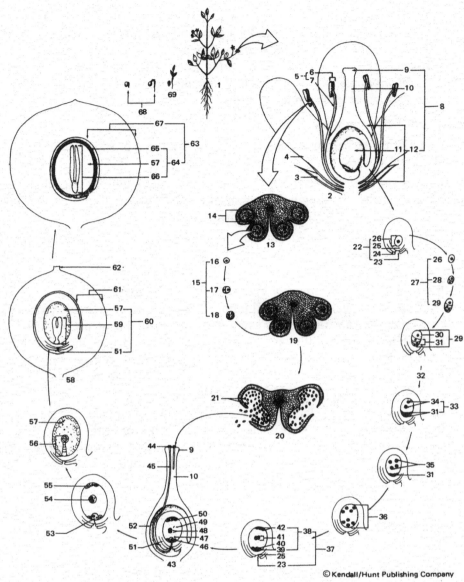

© Kendall/Hunt Publishing Company

1. MATURE SPOROPHYTE
2. FLOWER
   3. SEPAL
   4. PETAL
   5. STAMEN
      6. ANTHER
      7. FILAMENT
   8. PISTIL
      9. STIGMA
      10. STYLE
      11. OVULE
      12. OVARY
13. YOUNG ANTHER (C.S.)
14. MICROSPORANGIUM
15. *Meiosis*
   16. MICROSPORE MOTHER CELL
   17. Dyad
   18. Tetrad of Microspores
19. MATURING ANTHER (C.S.)
20. MATURE ANTHER (C.S.)
21. Pollen grains (male gametophyte)
22. YOUNG OVULE (L.S.)
   23. MICROPYLE
   24. INTEGUMENTS

25. MEGASPORANGIUM
26. MEGASPORE MOTHER CELL
27. *Meiosis*
   28. Dyad
   29. Linear tetrad of megaspores
30. Functional megaspore
31. Disintegrating megaspores
32. *Mitosis* (Functional megaspore)
33. Young embryo sac (female gametophyte)
34. Haploid nuclei
35. 4-Nucleate embryo sac
36. 8-Nucleate embryo sac
37. Mature ovule
38. Mature embryo sac
   39. Synergids
   40. Egg cell
   41. Polar nuclei
   42. Antipodal nuclei
43. *Double fertilization*
   44. Germinated pollen grain
   45. Pollen tube
   46. Egg cell
   47. Sperm nucleus
   48. Polar nuclei

49. Sperm nucleus
50. Antipodals
51. INTEGUMENTS OF OVULE
52. OVARY WALL
53. ZYGOTE *(After Fertilization)*
54. ENDOSPERM NUCLEUS (3N)
55. Deteriorating antipodals
56. YOUNG SPOROPHYTE EMBRYO
57. ENDOSPERM TISSUE (3N)
58. YOUNG FRUIT
   59. EMBRYO DIFFERENTIATED
   60. YOUNG SEED
   61. WALL OF FRUIT
62. REMNANTS OF STYLE
63. MATURE FRUIT
64. SEED
65. SEED COAT
66. EMBRYO
67. PERICARP (FRUIT WALL)
68. SEED GERMINATING
69. YOUNG SPOROPHYTE

Figure 25-3. Life cycle of an angiosperm.

III. Pollination and Fertilization

The transfer of pollen from anther to stigma is called pollination. Pollination can be accomplished by insects, wind and occasionally by birds, bats and water. The pollen transfer may be from an anther to a stigma that is genetically identical. This is termed *self-pollination*. Pollen transfer between two genetically different plants of the same species is termed *cross-pollination*. Since sperm are formed inside pollen, pollination precedes fertilization. When pollen comes into contact with the stigma, it begins to germinate forming a pollen tube that grows down the style to eventually enter the micropyle of an ovule inside the ovary. A sperm released from the pollen tube unites with the egg to form a fertilized egg or zygote.

Dust pollen grains from fresh flowers (or stored pollen) onto separate drops of water, 1%, 5%, 10% and 30% sucrose solutions on different coverslips. Obtain a depression slide for each coverslip and place a small amount of vaseline around the edge of the concavity with a toothpick. Invert the depression slide, press it against a coverslip and quickly upright it. Label each preparation and set it aside for at least 30 minutes. Later, observe it for pollen germination and the growth of pollen tubes. Describe the results below. Sketch a pollen grain which has produced a pollen tube.

Following fertilization, the zygote grows and differentiates to form an embryo. Integuments give rise to seed coats. In short, each ovule develops into a seed containing stored food and an embryo. The ovary enlarges forming a fruit with seeds on the inside. Both seeds and fruits are extremely important in food production for humans.

IV. The Seed (Figure 25–4)

Obtain and examine a soaked lima bean (dicot) seed. (Soaking initiates the process of seed germination and softens the seed coats.) Located in the center of the concave (indented) side is a scar, the *hilum*, which represents the point of attachment to a stalk connected to the placenta. Near the hilum, the micropyle is still evident as a small pore. Remove the seed coats! The remainder of the seed is the embryo.

The two *cotyledons* (cotyledon = seed leaf) associated with the embryo give *dicots* (short for dicotyledons) their name. Gently separate the cotyledons and locate the embryonic axis that is attached to both cotyledons at the cotyledonary node. The part of the embryonic axis above its point of attachment to the cotyledons is termed *epicotyl*. The part below the point of attachment is termed *hypocotyl*.

The epicotyl consists of a short portion of the stem including the stem apical meristem and a few small leaves. The hypocotyl consists of the embryonic stem below the cotyledonary node. The end of the hypocotyl gives rise to the root and is termed a *radicle*.

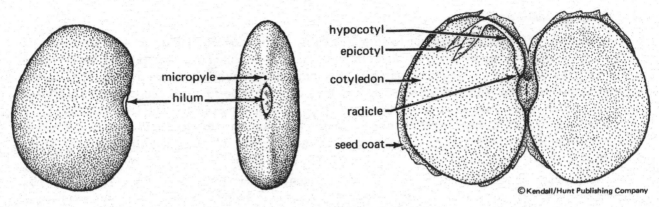

Figure 25-4. Structure of the seed.

V. The Fruit

A. Fruit Terminology.

Fruits may provide temporary protection for seeds and aid in their dispersal. Botanically, a fruit is defined as a matured ovary (or ovaries) together with any flower structures closely associated with it (them).

Terms describing fruits include:

1. *Simple*—developing from a single (simple or compound) pistil together with any other flower parts, i.e., green bean pod.

2. *Aggregate*—developing from several matured ovaries existing as a unit on a common receptacle together with any other flower parts, i.e., strawberry.

3. *Multiple*—developing from the ovaries of many flowers grouped into a single unit together with any other flower parts, i.e., pineapple.

4. *Dry*—the matured wall of the ovary, termed the *pericarp,* together with any other flower parts comprising the fruit are dry when mature, i.e., green bean pod.

5. *Fleshy*—part of the pericarp together with any other flower part comprising the fruit are fleshy, i.e., peach.

6. *Dehiscent*—splitting open at maturity, i.e., green bean pod.

7. *Indehiscent*—not splitting open at maturity, i.e., acorn.

B. Kinds of Fruits.

1. Simple dry dehiscent.

a. *Legume*—obtain a green bean pod (legume). Examine it. It is the fruit of the green bean plant. The pericarp is still relatively fresh and green. At maturity, the pericarp is usually dry and splits open along two lines to release the seeds. Break open the pod and locate the seeds. Note that each seed is attached by a short stock to the placenta. In addition, all of the seeds are located in a single compartment derived from a simple pistil. The fruits of peas and lima beans are also of this type.

b. *Capsule*—capsules develop from a compound pistil with the ovary having two or more compartments containing ovules. Ultimately, the fruit has two or more compartments containing seeds and dehisces at maturity. *Amaryllis, Datura* (jimson weed), *Ipomoea* (morning glory), *Iris* and poppies produce capsules. Examine available specimens.

2. Simple, dry, indehiscent.

   a. *Grains*—grains are one-seeded fruits. The pericarp is firmly joined to the seed coat over its whole surface and is not readily separated from it. Grasses (i.e., barley, corn, rice, rye and wheat) form this type of fruit. Examine available specimens.

   b. *Achene*—obtain a sunflower "seed." What is commonly called a sunflower seed is actually a one-seeded fruit termed an achene. The seed is attached to the pericarp at one point and the two are readily separated. Open the fruit and determine to which end of the fruit the seed is attached (broad end, narrow end). Other plants producing achenes are buck wheat and lettuce.

   c. *Samara*—the samara is a one- or two-seeded fruit similar to an achene but with a wing-like extension. Ashes, birches, elms, maples and boxelders produce samaras. Examine available specimens.

   d. *Nut*—the term nut as popularly used is often incorrect botanically. Peanuts are not nuts. They are seeds produced in pods (legumes). Coconuts are drupes to be described below. Brazil nuts are seeds. A true nut has a hard pericarp, contains one or occasionally two seeds and usually has an *involucre*. The involucre forms from specialized leaves that partially or entirely surround the mature nut. The cup of the acorn is an involucre. The husk of black walnut includes the involucre. Beechnuts, chestnuts, hazelnuts and hickory nuts are also true nuts. Examine available specimens.

3. Simple, fleshy

   a. *Drupe*—drupes develop from a simple pistil. Obtain an olive and examine it. The pericarp is divided into an outer thin skinlike exocarp, a fleshy mesocarp and an inner hard endocarp containing a single seed. Together the endocarp and seed constitute the pit or stone. Apricots, cherries, peaches, and plums are also drupes.

   b. *Berry*—berries develop from a compound pistil. The entire pericarp is fleshy with the exception of an outer skinlike exocarp. Examples include cucumbers, grapes, lemons, oranges, squashes and tomatoes. Examine available specimens.

4. Aggregate.
   Blackberries, strawberries and raspberries are aggregate fruits consisting of a number of similar small fruits. The individual fruits composing the aggregate fruit can be classified as simple fruits. Blackberries and raspberries consist of small drupes; strawberries of small achenes. Examine available specimens.

5. Multiple.
   Figs, mulberries, osage oranges and pineapples are multiple fruits formed by the matured ovaries of several flowers. The individual fruits composing the multiple fruit can be classified as simple fruits. Examine available specimens.

VI. Seedling Development

A viable seed placed under suitable environmental conditions germinates. The embryo it contains grows and differentiates to form a young plant or seedling.

A. Lima Bean Seedling Development (Figure 25–5).

Uproot and observe progressively older stages of lima bean (*Phaseolus lunatus*) seedling development to the point of formation of compound leaves. Which part of the embryonic axis

(epicotyl, hypocotyl, radicle) emerges from the seed first? _____

In some plants the cotyledon(s) together with the seed coat remain in the soil. Germination is said to be *hypogeous*. In others with *epigeous* germination the cotyledons are lifted above the

ground. Is germination of lima bean seeds epigeous or hypogeous? _____

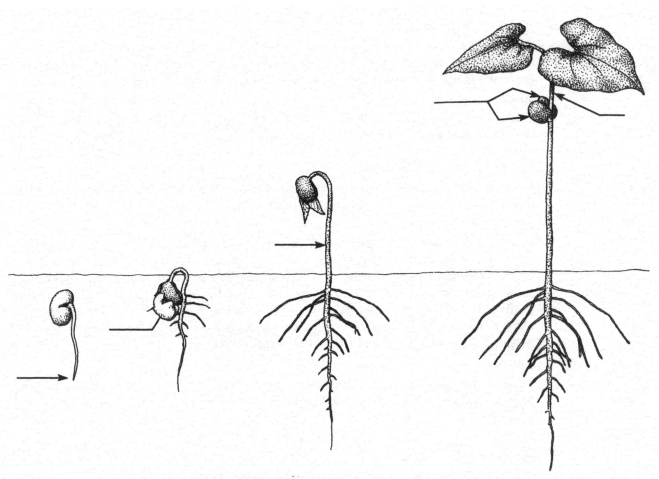

Figure 25-5. Seed and seedling of lima bean.

How many cotyledons are present? _____ What part of the embryonic axis (epi-

cotyl, hypocotyl, radicle) lifts the two cotyledons (beans are dicots) into the air? _____

The first structure to grow out from the seed coat is the radicle or embryonic root. It emerges through the micropyle to form the first root or primary root. Meanwhile the developing hypocotyl bends forming a hook or crozier. As the hypocotyl elongates the apex of the hook emerges from the soil with the cotyledons and epicotyl dragging after it. This arrangement affords protection to the cotyledons and epicotyl as they emerge from the soil. Eventually, the hypocotyl straightens lifting the cotyledons and epicotyl lightward.

Which part of the embryonic axis (epicotyl, hypocotyl, radicle) gives rise to new stem and foliage

leaves? _____ Are cotyledons present on older lima bean seedlings? _____

What color are the cotyledons that have been exposed to light? _____

The young cotyledons when exposed to light turn green and carry on photosynthesis. Only a small amount of food is made by them. Their stored food is quickly exhausted by the fast growing seedling. Within a short period of time the cotyledons shrivel up and drop off as newly produced foliage leaves assume their role in photosynthesis. Label radicle, seed coat, hypocotyl, epicotyl and cotyledons in Fig. 25-5.

B. Pea Seedling Development (Figure 25-6).

Uproot and observe progressively older stages of pea (*Pisum sativum*) seedling development. Is

germination of pea seeds epigeous or hypogeous? _____

302

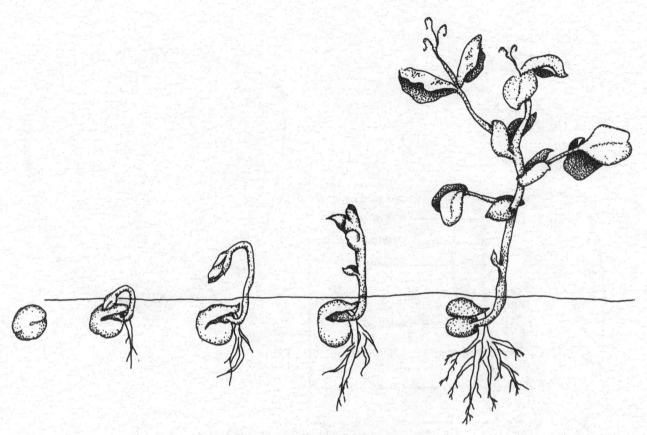

Figure 25–6. Seed and seedling of pea.

Although they remain below the ground, the cotyledons of pea seeds store food, as do those of lima bean seeds, for use during early stages of development. Which portion of the embryonic axis (epicotyl, hypocotyl) forms the bulk of the stem of the young pea seedling?

_____

How is the apical meristem (tip of the epicotyl) of the pea stem protected as it emerges from

the soil? Hint: Is a hook present? _____ If a hook is present, does it originate from

the epicotyl or hypocotyl? _____

C. Corn Seedling Development (Figure 25–7).

Uproot and observe progressively older stages of corn (*Zea mays*) seedling development. Recall that the corn grain is a fruit containing a single seed. Inside the seed coat is the embryonic axis with a single attached cotyledon (corn is a monocot), and a store of food called endosperm.

During germination, the cotyledon absorbs the stored endosperm for use by the embryo.

A tubular sheath called the *coleorhiza* surrounds the radicle. The embryonic root quickly pushes through the coleorhiza to form the primary root system. However, the primary root system never becomes very extensive. The main root system of the mature corn plant is adventitious in origin, developing from nodes located near the base of the stem. Similarly, the short apical meristem at the tip of the epicotyl is enveloped by the *coleoptile* sheath. It provides protection for the shoot apex as it emerges from the soil. Soon, the first leaf extends beyond the coleoptile and begins photosynthesizing.

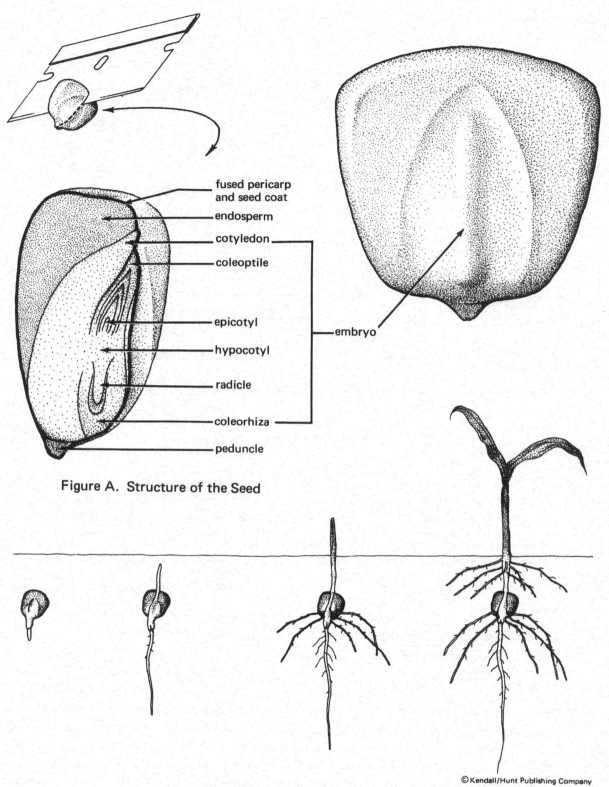

fused pericarp
and seed coat

endosperm

cotyledon

coleoptile

epicotyl

hypocotyl

radicle

coleorhiza

peduncle

embryo

Figure A. Structure of the Seed

© Kendall/Hunt Publishing Company

Figure 25-7. Seed and seedling of corn.

VII. Plant Movements

Plants, when contrasted with animals, are usually described as stationary. While this is true for the plant as a whole, plant parts may move in response to stimuli. Two types of plant movements in response to external stimuli are *tropisms* and *nastic movements*.

A. Tropisms.

Tropisms are responses where the direction of the movement of stem and root tips is determined by the direction from which the stimulus comes. Such movements result from unequal growth rates of cells that either turn the organ toward (positive tropism) or away from (negative tropism) the stimulus. These variations in cell growth are caused by plant hormones. The movement can be in response to light (phototropism), gravity (geotropism), touch (thigmotropism), water (hydrotropism), temperature (thermotropism), etc.

1. Phototropism.

Observe, describe and explain the differences between the stem tips of bean seedlings that have been uniformly lighted and those that have been exposed to light from only one side.

Does the stem tip exhibit positive or negative phototropism? _____

2. Geotropism.

Observe, describe and explain the differences between the stem tips of corn seedlings placed in different positions with respect to gravity. What type of tropism is represented? Is it negative or positive?

B. Nastic Movements.

In nastic movements the direction of movement bears no relationship to the direction from which the stimulus is applied. Such movements may occur in response to changes in light (photonasty), temperature (thermonasty), touch (thigmonasty), etc. The movements involve changes in turgor that increase or decrease the size of cells as they gain or lose water.

1. Thigmonasty.

Observe, describe and explain the response of the leaves of *Mimosa pudica,* "the sensitive plant," to touch.

Does the movement of *Mimosa* leaves result from unequal cell enlargement or from changes in turgor? _____

2. Describe an experimental procedure by which you could determine whether the opening and closing of morning glory flowers is in response to light, temperature or both. Be sure to include a hypothesis, materials required, and methods.

Plant Hormones

Plant hormones are substances produced by plant tissues for transport elsewhere in the plant to regulate various aspects of growth and development. At least five types of hormones have been described in flowering plants: auxins, gibberellins, cytokinins, abscisic acid and ethylene. Observe the following demonstrations.

A. Effect of a commercial rooting agent on root development—Compare *Coleus* cuttings treated with a commercial rooting agent containing auxins with untreated *Coleus* cuttings for the amount of root development. What effect do the auxins have on root development?

B. Effect of gibberellic acid on stem elongation and internode length—Compare stem and internode length on treated (the above ground portion of the plant is sprayed with 1,000 ppm gibberellic acid solution) and untreated pea seedlings or on treated and untreated dwarf seedlings.

C. Effects of removal of the stem apex on lateral bud development—Compare lateral bud development along *Coleus* stems that have had the stem tips removed with those that have not. How can you explain the observed difference?

REVIEW QUESTIONS

1. The male, pollen producing flower parts are called _____ .

2. The egg which is one cell of the embryo sac is located inside an ovule inside an _____ at the base of the pistil.

3. Sperm nuclei inside pollen are carried to the stigma by the process of _____

4. Growth of a _____ down through the style transports sperm close to the egg inside the ovule.

5. The pollen tube enters the ovule by growing through an opening called the _____

6. Following fertilization, the ovule matures into a _____ containing an embryonic plant.

7. The ovary matures into a _____ .

8. The fertilization of polar nuclei by a sperm produces _____ .

9. The green bean fruit is termed a _____ .

10. Since it opens at maturity, the green bean fruit is said to be _____ .

11. The portion of the embryonic axis below its point of attachment to cotyledons is termed _____

12. The structure that protects the stem apical meristem as it emerges from the soil is the _____

13. A type of plant movement in which the direction of the movement is determined by the direction from which the stimulus comes. _____ .

14. The five main types of hormones in flowering plants are:

    a. _____   b. _____   c. _____

    d. _____   e. _____

POND ECOLOGY

Objectives

Upon completion of this exercise you should be able to:

1. Explain the importance of water to living organisms in aquatic and nearby habitats.
2. Define ecology, ecosystem, community, population, trophic level, producer, consumer, decomposer, habitat and niche.
3. Define succession and briefly describe the major events which lead to the replacement of a pond by a forest community.
4. Describe the significance of the first and second laws of thermodynamics to the entry, flow and utilization of energy in an ecosystem.
5. Diagram a profile of the pond in which you indicate the organisms in each zone.
6. Describe how physical factors of the environment affect the distribution, diversity and abundance of organisms in the pond.
7. Answer all questions posed in the exercise.

I. Introduction

Ecology is the study of the interrelationships between living organisms and their environment. The individuals of the same species, in a given area, are referred to as a *population*. A *community* consists of different populations of different species that live together in the same environment or habitat. The term *ecosystem* is used to refer to the living (community) and the nonliving components of the environment. The capturing of energy, the use of energy, and the transfer of energy constitute the heart of ecological relationships. Energy transfer is governed by the physical laws of thermodynamics. The First Law of Thermodynamics briefly says that energy is neither created nor destroyed, but only changed in form. The Second Law of Thermodynamics says that when energy is transferred there is a reduction in usable energy.

In order to demonstrate the dynamic aspect of ecology, you will be involved in a study of a specialized community, the pond. During the course of this lab you will be called upon to make many observations and draw conclusions from them.

II. Water

Excluding energy, life in aquatic communities is largely dependent upon the unique properties of water. Before lab, study the water cycle and answer the questions in this section about the properties of water.

A. Water Cycle

Water is continuously cycling between the land and atmosphere. *Evaporation* or the conversion of liquid water to a vapor is powered by the energy of sunlight. As water vapor enters the atmosphere it rises, cools, condenses, and returns to the land as *precipitation* (rain or snow).

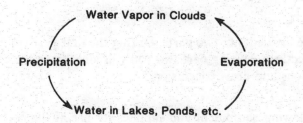

B. Properties of Water

   1. Describe the polar (dipole) molecule and hydrogen bonding.

   2. Explain why water is a good solvent and transport medium for chemical nutrients in ecosystems.

   3. Define surface tension and how it aids the movement of some organisms on the surface of water.

   4. Define density and describe how the density of water changes in relation to temperature. Explain the ecological importance.

   5. Define specific heat, heat of fusion and heat of vaporization. Explain how these help to moderate daily and seasonal fluctuation in the temperature of aquatic and nearby habitats.

III. Laboratory Procedures

The laboratory will be divided into two parts. During the first part you will go to the pond and make observations and collections. During part two you will identify collected organisms, analyze and organize your observations and draw conclusions.

Part 1. To the Pond:

You will need the following equipment and supplies for use at the pond:

      collecting jars (12 to 15)
      plankton nets (2)
      portable oxygen and temperature meter
      thermometers (2)
      secchi disk
      fish seine
      weighted line marked off in meters
      boat, oars and life preservers
      field guides
      portable pH meter or pH paper
      other equipment as indicated by your instructor

(Note: Safe practices and care should be exercised. Your instructor will brief you about any potential hazards.)

Following is a list of specific observations and collections that need to be made:

A. Mapping the Pond

  1. Draw an outline of the pond and show the points at which water enters and exits and the surrounding vegetation. A very serviceable map can be constructed using the method described by Nutting (Amer. Bio. Teacher, 28(5):351–360).

  2. Set up transect lines at about 10 to 15 meter intervals. Depth recordings can be made and recorded for several points along each transect by lowering a weighted, marked line from a boat. Finally, contour lines can be drawn by joining points of equal depth.

  3. Information about the rooted pond vegetation can be included as obtained.

B. Succession

The pond is a dynamic community. It changes with time. These changes are referred to ecologically as succession. In a totally natural setting (with no interference from humans) a pond will slowly fill in and become more and more shallow. A typical succession in eastern North America would be:

pond --------→wet meadow --------→dry meadow --------→forest.
             (swampy)   (field)

There are many factors involved in succession. Hopefully, you will be able to determine some of the physical and biotic factors during the lab.

  1. What evidence indicates that the pond is undergoing succession?

2. What physical and biological factors are causing the pond to become more shallow?

3. Identify and list the types of animals and rooted plants that you see beginning on dry ground at the edge of the pond and proceeding into the water. Indicate the relative abundance of each as few (+) to many (+++++). Consult appropriate field guides to identify the organisms observed.

Plants Observed and Where   **Animals Observed and Where**

C. Pond Or Lake?

A pond is usually defined as a shallow, small body of water having rooted plants over most of its bottom. Beginning at the shoreline three zones of rooted plants may occur: emergent, floating and submerged.

1. Is the body of water you are studying a lake or pond? Explain your answer.

2. Are emergent, floating and submerged plants present? If so list them.

Emergent Plants   **Floating Plants**   **Submerged Plants**

D. Turbidity

Turbidity or the degree of cloudiness determines the depth to which effective light penetration occurs. The depth at which the energy utilization in respiration equals the energy production in photosynthesis is referred to as the *compensation point*. An estimate of the compensation point can be made as follows:

1. Lower a secchi disk and record the depth in meters at which it disappears.

2. After dropping the secchi disk for two more meters, if possible, raise it and record the depth at which the disk reappears.

Compensation $\approx$ (Recording 1 + Recording 2) / 2
point   (_____ m.)   (_____m.)

        $\approx$ _____ m.

E. pH

Determine and record the pH of the water at each of the following locations:

Location	pH
Emergent plant zone	_____
Submerged plant zone	_____
Deep water zone (no rooted plants)	_____

F. Temperature

1. Determine and record the temperature at each of the following locations:

Location	Temperature (°C)
Air	_____
Surface of ground (dry zone)	_____
Emergent zone	_____
Submerged zone	_____
Deep water zone	_____

2. Determine and record the water temperature at different depths in the deep water zone with a temperature meter. Alternatively, collect water at different depths and determine the temperature with a thermometer as soon as each sample is collected.

Depth (m.)   **Temperature (°C)**

G. Dissolved Oxygen

1. Determine the dissolved oxygen content in milligrams per liter (mg/1) at the locations listed below using a portable dissolved oxygen meter.

Location	Dissolved Oxygen Content
Emergent zone	_____
Submerged zone	_____
Deep water zone	_____

2. Determine the dissolved oxygen content at different depths in the deep water zone.

Depth   **Dissolved Oxygen Content**

H. Pond Life

While abundance, distribution and diversity of organisms varies, energy enters and flows through at least three categories of organisms: producers, consumers and decomposers. Energy can be viewed as flowing from one trophic (feeding) level to another with less than 100% efficiency. *Producers* (photosynthesizing organisms mainly) are the first trophic level. The sugar made by them in photosynthesis is used by them or is stored. All other organisms depend on producers directly or indirectly for their energy. *Primary consumers* (herbivores) are at the second trophic level and feed on producers. *Secondary consumers* (carnivores) are at the third trophic level and eat herbivores. Carnivores may be consumed by other carnivores (tertiary consumers) occupying a fourth trophic level. *Decomposers,* bacteria and fungi, decompose dead organisms from all trophic levels providing a means for recycling chemical elements through the ecosystem. Energy is not recycled. It must be continually replenished by the sun (recall the first and second laws of thermodynamics).

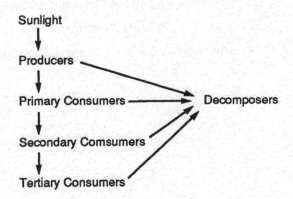

The tiny free floating, drifting and weak swimming organisms in open waters are termed *plankton.* Planktonic producers (blue green bacteria, diatoms, desmids, etc.) are called *phytoplankton.* The planktonic consumers (protozoa, rotifers, cladocerans, copepods, etc.) are called *zooplankton.*

1. Define producer, consumer, and decomposer.

2. Collect three plankton samples from the surface water of the deep water zone by pulling a plankton net rapidly through the water. Place each sample in a separate jar labeled "plankton-surface water sample." Also, collect three plankton samples from deeper water at the same location by pulling the plankton net slowly through the water. This technique permits the net to sink. Place each sample in a separate jar labeled "plankton-deep water sample."

3. Weather permitting, sample the emergent plant zone and the submerged plant zone with a fish seine by pulling it through the water. Determine the numbers and kinds of animals in each zone.

Animals Present   Number

4. Collect an undiluted, unconcentrated water sample in the deep water zone by dipping a clean jar in the water. Label the jar "pond water sample."

5. Collect about a cup full of mud from each of the following locations: emergent plant zone, submerged plant zone and deep water zone. Place each sample in a separate labeled jar.

Part 2. To the Lab:

A. When you get back to the laboratory make sure you wash all of the collecting equipment that you used.

B. Observe each plankton sample under the dissecting and compound microscopes. Use the available field guides to identify planktonic organisms. Categorize each different kind of organism as producer or consumer and indicate its relative abundance (+ to + + + +). A more precise assessment of the phytoplankton can be made using the method described by Sohn (Amer. Bio. Teacher, 34(1):19–22) involving filtering a known volume of pond water through a membrane filter and counting the collected cells with the aid of a microscope.

Surface Samples

Name of Producer	Relative Abundance	Name of Consumer	Relative Abundance

Deep Water Samples

Name of Producer	Relative Abundance	Name of Consumer	Relative Abundance

C. Place each mud sample in a sieve and observe it for living organisms. Identify all organisms with the aid of a dissecting microscope and keys, and indicate their relative abundance.

Emergent Zone		Submerged Zone		Deep Water Zone	
Organism	Relative Abundance	Organism	Relative Abundance	Organism	Relative Abundance

D. Use the standard pour plate technique to determine the number of bacteria per milliliter of the pond water sample. Refer to any introductory microbiology laboratory manual for the procedure.

REVIEW QUESTIONS

1. Now that you have completed your observations on the pond ecosystem classify the organisms observed as occupying the first, second or third trophic level.

First Trophic Level Organisms	Relative Abundance	Second Trophic Level Organisms	Relative Abundance	Third Trophic Level Organisms	Relative Abundance

2. At which trophic level(s) are organisms most abundant and why?

3. Diagram a food chain that is evident from your data.

4. Describe at least three ways that the aquatic ecosystem interacts with or is linked to the surrounding terrestrial ecosystem.

5. What "conclusions" can you make from the depth, turbidity, pH, temperature, dissolved oxygen and producer data? If data are collected at regular intervals throughout the year, discuss the seasonal changes that occur in the pond ecosystem.

6. The *habitat* of an organism can be described as its "home address" and its *niche* as its "occupation." For each of the following organisms list the habitat and niche.

   a. Willow tree

   b. Diatom

   c. Hydra

   d. Frog

7. Diagram an energy pyramid and explain it.

8. Describe the main events that cause succession to take place in a pond.

9. Complete each of the following statements by filling in the blanks.

   a. The term _____ refers to the biological community and its physical environment.

   b. The temperature at which water possesses its greatest density is _____ °C.

   c. The _____ category of organisms is responsible for capturing and making energy available to other organisms in the ecosystem.

   d. The main producers in the surface waters of the deep water zone are the _____ .

   e. The depth at which photosynthesis equals respiration is referred to as the _____ .

   f. _____ is the environmental factor that determines the maximum depth at which phytoplankton can exist.

SYMBIOSIS

Objectives

Upon completion of this exercise you should be able to:

1. Define symbiosis.
2. Define the three main types of symbiotic relationships and give examples of each.
3. Distinguish between predator-prey and symbiotic relationships.
4. Recognize and name anatomical structures and modifications of flukes and tapeworms which are adaptations to a parasitic way of life.
5. Answer all review questions.

I. Introduction

Symbiosis refers to an intimate relationship between two organisms of different species who live in, on, or in contact with each other. The basis for the relationship is usually food. The word symbiosis comes from two words that mean "living together." Some form of symbiotic association can be found in every phylum of organisms.

Predation is another type of relationship between organisms, but it is quite different. Predation is not intimate. The predator is typically the larger organism who eats the smaller organism, the prey. Some examples of predator and prey are: the lion and the gazelle; shark and flounder, cat and mouse; praying mantis and fly. Even herbivores can be considered to be predators on the plants they eat.

Usually three levels of symbiosis are recognized: mutualism, commensalism, and parasitism.

A. Mutualism—both members of the association benefit. For example humans have bacteria by the name of *Escherichia coli* in their lower digestive tract. The bacteria derive their food from undigested material. In their metabolism they make such products as vitamin B_{12}, vitamin K, and niacin which are absorbed and used by the human who cannot synthesize them. Thus, both members benefit. Mutualism is fairly common.

B. Commensalism—one species receives some benefit from the association but the other species is neither harmed nor benefited. Hundreds of species of bacteria are known to live in the human colon deriving their food from the human host, but they cause no damage nor are they known to contribute anything to man. When commensalistic associations are examined closely they often prove to be beneficial to both or harmful to one and therefore not truly commensal. Commensalism is not common.

C. Parasitism—one species (the parasite) lives by deriving its food of body fluids, cells or cell products from the other species (the host). Clearly one is harmed and the parasite benefits. Parasites you may be familiar with are tapeworms, lice, and the bacteria that cause syphilis, tuberculosis and whooping cough. Parasitism is very common.

Parasites may fall into one of several classes.

   1. *Ectoparasites,* which may be found on the surfaces of organisms (e.g. fleas),

   2. *Endoparasites,* which live within the bodies of their hosts (e.g. tapeworms, flukes).

      a. Intracellular endoparasites live within the cells of their hosts (e.g., some bacteria, malaria organisms)

b. Extracellular endoparasites live outside of the cells of their hosts (most bacteria and parasitic worms fit this description).

Parasites may also be obligate (dependent upon very specific hosts), or facultative (capable of a free existence).

View the film on the parasitic flatworms and pay attention to the modifications of the body anatomy, physiology, and life cycle which adapt a flatworm to the parasitic way of life.

II. Examples of Associations Between Organisms

1. *Mycorrhiza* is a symbiotic association between a fungus and the roots of a vascular plant. The fungus involved may be either a single-celled type which lives within the cells of the root or it may be a multicellular type which covers the root tip with mats of hyphae, some of whose branches extend between cells of the root cortex.

   Plants with mycorrhizae take up water and some essential nutrients (e.g. phosphate, potassium and calcium) better than those without; some plants can, in fact, not survive in the absence of mycorrhizae. Experiments with radioactively-labeled carbon have demonstrated that sugars manufactured by the plant are taken up by the fungus for its own nourishment. Obtain the slide labeled "Endotrophic Mycorrhiza", which shows a cross-section of an orchid root. Orchid seeds will not develop in nature unless they are infected with specific species of fungi. The fungi accumulate nutrients and vitamins and contribute them to the orchid. In return, the fungus takes up permanent residence in the roots of the orchid, living *within* the cortex cells. The fungus receives organic nutrients from the orchid, while apparently aiding the orchid in the accumulation of inorganic materials.

   Under low power, cells infected with fungi appear filled with reddish or bluish spaghetti, while uninfected orchid cells have nuclei and prominent starch grains as their only visible contents. If you switch to high power, the "spaghetti" resolves into tangled masses of fungal hyphae. Sometimes the fungus grows so vigorously that the orchid cell is killed, while in other cases both organisms cohabit the same cell.

2. *Termites and Their Flagellates.* Termites are insects which eat wood but which cannot digest cellulose themselves. They depend on a cellulose-digesting enzyme produced within flagellated protozoans, which live in the termite intestine. Termites chew and ingest wood, which is then digested by the flagellate. Both animals use the glucose produced by the breakdown of cellulose.

   If live specimens of termites are available, obtain one of them. After examining its large mandibles adapted for tearing wood, remove a section of its abdomen and mount it in 0.4% saline. Look for the flagellated protozoan. Sketch one or more of the flagellates.

   If available examine a prepared slide labeled "termite flagellates w.m." and compare it with your sketch.

3. *Lichens.* A lichen is a mutualistic partnership between an alga or cyanobacterium and a fungus. The fungus is usually an ascomycete or occasionally a basidiomycete. Together the two members of the partnership make a successful combination capable of growing on rocks, soil or trees. They form an important part of the vegetation in the tundra biome and are often "pioneer plants" involved in the early colonization of harsh, exposed surface (i.e., bare rocks).

Obtain a prepared slide of a lichen and identify the colorless fungal hyphae and the red stained algal cells.

Three growth forms occur among species of lichens. They are: *Crustose* type—This is thought to be the most primitive type. The body is flat and firmly attached to the surface below. Only small pieces can be removed from the bark, rock or soil; *Foliose* type—The body is flat, leaflike, branched and loosely attached to bark, rocks or soil; *Fruticose* type—the body is erect and branching.

Examine representative species on display and classify them according to growth form.

	Species	**Growth Form**
1.	_____	_____
2.	_____	_____
3.	_____	_____

4. *Chinese liver fluke.* This worm infects about 20 million humans in the Orient where people eat raw fish. The life cycle of the parasite involves three hosts; a fresh water snail, a fish and man. Humans become infected by eating the infective cyst stage in raw or undercooked fish. The proper snail host does not occur in the United States and therefore this fluke does not occur here. All the members of the class Trematoda of the flatworm phylum are parasitic. They are not capable of synthesizing sterols (a type of lipid) or fatty acids and this may account for their dependence on a host.

Examine the slide *Clonorchis sinensis* w.m. under low power. This Chinese liver fluke possesses an anterior *oral sucker* and a large *ventral sucker.* Beginning at the oral sucker trace the digestive tract past the short muscular *pharynx* and forked *gastrovascular cavity.*

The reproductive system consists of two *testes* located at the posterior of the organism. Sperm produced here travel through a duct to the genital pore located adjacent to the ventral sucker. Eggs are produced in a lobed *ovary* which is located anterior to the testes. In front of the ovary is a large convoluted duct (the *uterus*) which serves as an egg storage area. Posterior to the ovary is the rounded seminal receptacle where sperm are stored.

Figure 27-1.

On Figure 27-1, label oral sucker, ventral sucker, pharynx, gastrovascular cavity, testes, ovary and uterus.

Note that in the flukes the reproductive system is quite elaborate. Selection has favored this direction since most of these parasites must complete their life cycle in two or more host organisms. Emphasis on reproduction increases the chances that at least one of the thousands of eggs fertilized will complete the journey to the next host.

Another specialization is demonstrated by the presence of muscular suckers which enable the parasite to attach to its host and to obtain vital nutrients from the host's tissues. Flukes possess a thickened epidermis without cilia which protects them from digestion by the host's enzymes.

5. *Tapeworm*

Examine the slide *Taenia pisiformis*. Like the fluke, the tapeworm has become adapted to the parasitic way of life. Observe the head region (scolex) and identify *suckers* and *hooks*. Posterior to the head and neck region is a long ribbon like series of sections *(proglottids)*. The neck region represents a growing area. Proglottids near this region are immature and contain only the male sex organs while proglottids further down the line are fully mature and contain both male and female sex organs. At the extreme distal end are the egg-filled "ripe" sections which are ready to be detached and pass out of the host.

Observe a mature section and identify:

a. testes: very small and scattered throughout the section.

b. ovaries: two dark-staining masses at the distal end of the proglottid.

c. uterus: slender tube running along midline of proglottid.

Now observe a ripe section. The main structure seen here is a greatly enlarged uterus filled with fertilized eggs.

The adult tapeworm inhabits the intestines of cats, dogs or foxes. The ripe proglottid passes out of the intestines in the feces. In order for the worm to complete the next phase of its life cycle the egg must be eaten by a rabbit. The probability of this occurrence is rather small. The tapeworm has adapted to this predicament in terms of its capability to produce thousands of eggs in each ripe proglottid.

Notice that none of the proglottids examined contained any digestive organs. How, then, does the tapeworm obtain its nutrients? _____

_____

By living in the intestine, the tapeworm relies on the host organism to obtain and digest the food. The worm then merely absorbs the digested material directly into the proglottids.

In Figure 27-2, label hooks, suckers, proglottids, testes, ovary, uterus.

Observe the demonstration of parasitic flatworms (if available).

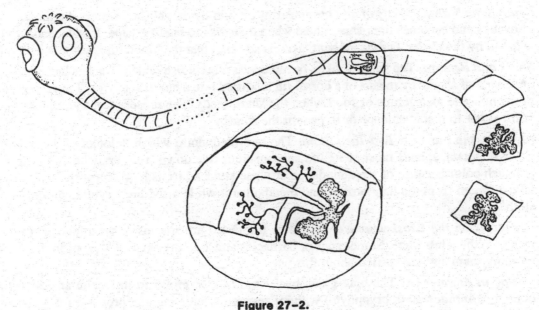

Figure 27-2.

6. Apicomplexa

All apicomplexans are parasites. Four species in the genus *Plasmodium* are known to cause malaria in humans. A sporozite stage is transmitted to humans by the bite of an infective female *Anopheles* mosquito. Subsequently, a repeating stage is established that occurs inside red blood cells.

Obtain a blood smear and identify a ring shaped stage of *Plasmodium* inside a red blood cell.

III. Analyze Each Association that Follows

For the following relationships between organisms, read the information given and decide whether it is a predator-prey association or a symbiotic one. If it is symbiotic is it mutualistic, commensalistic or parasitic? Ectoparasitic or endoparasitic? Obligate or facultative? Record your answers and list the advantages received or harm suffered by each member of the association. Demonstrations of some of these interactions may be available in lab.

1. *Chlorohydra.* These "green hydra" are dependent for their nutrition on green algae (zoochlorellae) which live and photosynthesize within their gastrodermal layer. The algal cells presumably benefit from the waste products of the hydra: carbon dioxide and nitrogenous excreta.

   If specimens of green hydra are available in the laboratory, mount one of these in a concavity slide in distilled water, cover with a coverslip, and observe under the microscope.

2. *Bacteria in Root Nodules.* The primary source of nitrogen incorporated into the bodies of living organisms is gaseous nitrogen from the atmosphere. Nitrogen cannot be used by either plants or animals in this form, however. It must be converted into nitrites and nitrates, which can be used by plants and passed along to animals. The first step in this conversion process is the reduction of N_2 to ammonia by the so-called "nitrogen-fixing" bacteria. Certain of these bacteria which belong to the genus *Rhizobium* live symbiotically in the roots of various kinds of plants, especially those, such as alfalfa, which belong to the bean family. These bacteria invade the inner cells of the root cortex, causing them to swell up; and clusters of such swollen cells form lumps, or nodules, on the root. The bacteria in the nodules are able to pick up nitrogen from the soil and reduce it to ammonia, which can then be used by the plant for making amino acids.

3. *Crab in the Mantle Cavity of an Oyster.* A tiny crab in the larval stage enters the mouth cavities of oysters and becomes imprisoned there when it grows too big to escape. It is dependent on food taken in by the oyster, but the amount taken is so small that the loss to the oyster is insignificant.

4. *Shark and Remora.* The shark-sucker, or remora, is a bony fish which attaches harmlessly to the underside of sharks by means of a dorsal fin modified into a holdfast device. The remora secures protection and also scraps of food when the shark feeds. The presence of the remora does not seem either to inconvenience or to benefit the shark.

5. *Algae on the Fur of a Two-Toed Sloth.* The two-toed sloth, which belongs to the mammalian order Edentata, spends most of its life suspended upside down from branches. The sloth has a greenish color because of the abundance of algae attached to its hair. Apparently the algae can take nitrogen from the hair surface and metabolize it while providing cryptic coloration for the sloth.

6. *Barnacles on the Exoskeletons of Lobsters.* Barnacles are species of arthropod which are not motile in the adult stage. Some types of barnacles attach to the exoskeletons of lobsters without harming them and are carried about this way.

7. *Nostoc in Anthoceros.* The blue-green bacterium *Nostoc* grows in intercellular cavities of the hornwort *Anthoceros* (a bryophyte) without apparently creating any problem for its host. Most blue-green bacteria can fix nitrogen.

8. *Anemone and Clown Fish*. Sea anemones regularly feed on fish which they paralyze by means of stinging cells in the tentacles. The small "clown fish", however, can live among the tentacles of sea anemones, and even occasionally enter the gastrovascular cavity of the anemone without coming to any harm. They achieve their immunity by swimming back and forth close enough to become covered with slime from the coelenterate, which then "recognizes" the fish as "self". The fish feed on scraps of food left over from the anemone's dinner.

9. *Plasmodium and man*. *Plasmodium* is a protozoan which lives inside the red blood cells of humans. While in the erythrocytes the protozoan eats hemoglobin and reproduces asexually. After two to three days the red blood cell bursts and 8 to 16 new *Plasmodium* organisms invade more blood cells. In man this causes the disease malaria which is characterized by a two to three day fever cycle and anemia. Mosquitoes of the genus *Anopheles* transmit the protozoan from human to human. Over 400 million humans suffer from malaria infections annually.

10. *Lamprey and fish*. The lamprey, one of the jawless fishes, attaches to a fish by its mouth. It rasps away tissue and eats the tissue and body fluids of the fish. Sometimes the fish dies and sometimes the lamprey drops off after several days and the fish survives.

ANSWER SHEET

1. _____

Hydra:

Algae:

2. _____

Bacteria:

Plant:

3. _____

Crab:

Oyster:

4. _____

Shark:

Remora:

5. _____

   Sloth:

   Algae:

6. _____

   Barnacle:

   Lobster:

7. _____

   Nostoc:

   Anthoceros:

8. _____

   Anemone:

   Clownfish:

9. _____

   Plasmodium:

   Human:

10. _____

   Lamprey:

   Fish:

REVIEW QUESTIONS

1. In the termite-flagellate association which member has the enzyme cellulase?

   _____

2. Cite the function for the

   a. scolex of a tapeworm

   b. suckers of a fluke

   c. seminal receptacle of a fluke

3. Is it always easy to categorize the associations between organisms?

4. The relationship between a woman and her unborn fetus is

   a. mutualism
   b. commensalism
   c. parasitism
   d. none of the above

5. Identify three ways flukes and tapeworms are specialized for their parasitic way of life.

6. In what ways is the tapeworm more specialized as a parasite than a fluke?

7. Match each organism listed below with one of the classes of parasites listed on the right.

   ____ Malarial protozoan (*Plasmodium*)   A. ectoparasite

   ____ Mosquito   B. extracellular endoparasite

   ____ Flea   C. intracellular endoparasite

   ____ Tapeworm

   ____ Tick

   ____ Rocky Mountain spotted fever bacterium

Periodic Chart of the Elements

Fisher Scientific

302C

Group (New IUPAC)	1	2	3	4	5	6	7	8	9	10	11	12	13	14	15	16	17	18
Former IUPAC	IA	IIA	IIIA	IVA	VA	VIA	VIIA	VIIIA	VIIIA	VIIIA	IB	IIB	IIIB	IVB	VB	VIB	VIIB	VIIIB
Former CAS	IA	IIA	IIIB	IVB	VB	VIB	VIIB	VIIIB	VIIIB	VIIIB	IB	IIB	IIIA	IVA	VA	VIA	VIIA	0

Row																		
1	H 1 1.00794$^\triangle$																	He 2 4.002602*
2	Li 3 6.941*	Be 4 9.01218											B 5 10.811$^\triangle$	C 6 12.011	N 7 14.0067	O 8 15.9994$^\triangle$	F 9 18.998403	Ne 10 20.179
3	Na 11 22.98977	Mg 12 24.305											Al 13 26.98154	Si 14 28.0855†	P 15 30.97376	S 16 32.066†	Cl 17 35.453	Ar 18 39.948
4	K 19 39.0983	Ca 20 40.078$^\triangle$	Sc 21 44.95591	Ti 22 47.88†	V 23 50.9415	Cr 24 51.9961$^\triangle$	Mn 25 54.9380	Fe 26 55.847†	Co 27 58.9332	Ni 28 58.69	Cu 29 63.546†	Zn 30 65.39*	Ga 31 69.723$^\triangle$	Ge 32 72.59†	As 33 74.9216	Se 34 78.96	Br 35 79.904	Kr 36 83.80
5	Rb 37 85.4678†	Sr 38 87.62	Y 39 88.9059	Zr 40 91.224*	Nb 41 92.9064	Mo 42 95.94	Tc 43 (98)	Ru 44 101.07*	Rh 45 102.9055	Pd 46 106.42	Ag 47 107.8682†	Cd 48 112.41	In 49 114.82	Sn 50 118.710$^\triangle$	Sb 51 121.75†	Te 52 127.60†	I 53 126.9045	Xe 54 131.29†
6	Cs 55 132.9054	Ba 56 137.33	**La 57 138.9055†	Hf 72 178.49†	Ta 73 180.9479	W 74 183.85†	Re 75 186.207	Os 76 190.2	Ir 77 192.22†	Pt 78 195.08†	Au 79 196.9665	Hg 80 200.59†	Tl 81 204.383	Pb 82 207.2	Bi 83 208.9804	Po 84 (209)	At 85 (210)	Rn 86 (222)
7	Fr 87 (223)	Ra 88 226.0254	▼Ac 89 227.0278	Unq 104 (261)	Unp 105 (262)	Unh 106 (263)												

**Lanthanides

Ce 58 140.12	Pr 59 140.9077	Nd 60 144.24†	Pm 61 (145)	Sm 62 150.36†	Eu 63 151.96	Gd 64 157.25†	Tb 65 158.9254	Dy 66 162.50†	Ho 67 164.9304	Er 68 167.26†	Tm 69 168.9342	Yb 70 173.04†	Lu 71 174.967

▼Actinides

Th 90 232.0381	Pa 91 231.0359	U 92 238.0289†	Np 93 237.0482	Pu 94 (244)	Am 95 (243)	Cm 96 (247)	Bk 97 (247)	Cf 98 (251)	Es 99 (252)	Fm 100 (257)	Md 101 (258)	No 102 (259)	Lr 103 (260)

- ● New IUPAC
- ■ Former IUPAC
- ◆ New Chemical Abstract Service
- ★ Former Chemical Abstract Service

FISHER SCIENTIFIC
CAT NO 05-702-10

§The International Union of Pure and Applied Chemistry (IUPAC) has not adopted official names or symbols for these elements

*These weights are considered reliable to ±2 in the last place.

†These weights are considered reliable to ±3 in the last place.

$^\triangle$These weights are considered reliable in the last place, as follows: Calcium and Gallium ±4; Boron ±5; Chromium and Sulfur ±6; Hydrogen and Tin ±7.

All other weights are reliable to ±1 in the last place. All reliabilities are based on an uncertainty scale of ±1 to 9

Atomic weights corrected to conform to the most recent values of the Commission on Atomic Weights. Column nomenclature conforms to IUPAC system and data in this chart have been checked by the National Bureau of Standards' Office of Standard Reference Data.

© 1987 Fisher Scientific

Atomic Weights

Name	Symbol	Atomic Number	Atomic Weight
Actinium	Ac	89	227.0278
Aluminum	Al	13	26.98154
Americium	Am	95	(243)
Antimony (Stibium)	Sb	51	121.75†
Argon	Ar	18	39.948
Arsenic	As	33	74.9216
Astatine	At	85	(210)
Barium	Ba	56	137.33
Berkelium	Bk	97	(247)
Beryllium	Be	4	9.01218
Bismuth	Bi	83	208.9804
Boron	B	5	10.811⌐
Bromine	Br	35	79.904
Cadmium	Cd	48	112.41
Caesium	Cs	55	132.9054
Calcium	Ca	20	40.078⌐
Californium	Cf	98	(251)
Carbon	C	6	12.011
Cerium	Ce	58	140.12
Chlorine	Cl	17	35.453
Chromium	Cr	24	51.9961⌐
Cobalt	Co	27	58.9332
Copper	Cu	29	63.546†
Curium	Cm	96	(247)
Dysprosium	Dy	66	162.50†
Einsteinium	Es	99	(252)
Erbium	Er	68	167.26†
Europium	Eu	63	151.96
Fermium	Fm	100	(257)
Fluorine	F	9	18.998403
Francium	Fr	87	(223)
Gadolinium	Gd	64	157.25†
Gallium	Ga	31	69.723⌐
Germanium	Ge	32	72.59†
Gold	Au	79	196.9665
Hafnium	Hf	72	178.49†
Helium	He	2	4.002602*
Holmium	Ho	67	164.9304
Hydrogen	H	1	1.00794⌐
Indium	In	49	114.82
Iodine	I	53	126.9045
Iridium	Ir	77	192.22†
Iron	Fe	26	55.847†
Krypton	Kr	36	83.80
Lanthanum	La	57	138.9055†
Lawrencium	Lr	103	(260)
Lead	Pb	82	207.2
Lithium	Li	3	6.941†
Lutetium	Lu	71	174.967
Magnesium	Mg	12	24.305
Manganese	Mn	25	54.9380
Mendelevium	Md	101	(258)
Mercury	Hg	80	200.59†
Molybdenum	Mo	42	95.94
Neodymium	Nd	60	144.24†
Neon	Ne	10	20.179
Neptunium	Np	93	237.0482
Nickel	Ni	28	58.69
Niobium	Nb	41	92.9064
Nitrogen	N	7	14.0067
Nobelium	No	102	(259)
Osmium	Os	76	190.2
Oxygen	O	8	15.9994†
Palladium	Pd	46	106.42
Phosphorus	P	15	30.97376
Platinum	Pt	78	195.08†
Plutonium	Pu	94	(244)
Polonium	Po	84	(209)
Potassium (Kalium)	K	19	39.0983
Praseodymium	Pr	59	140.9077
Promethium	Pm	61	(145)
Protactinium	Pa	91	231.0359
Radium	Ra	88	226.0254
Radon	Rn	86	(222)
Rhenium	Re	75	186.207
Rhodium	Rh	45	102.9055
Rubidium	Rb	37	85.4678†
Ruthenium	Ru	44	101.07†
Samarium	Sm	62	150.36†
Scandium	Sc	21	44.95591
Selenium	Se	34	78.96†
Silicon	Si	14	28.0855†
Silver	Ag	47	107.8682
Sodium (Natrium)	Na	11	22.98977
Strontium	Sr	38	87.62
Sulfur	S	16	32.066⌐
Tantalum	Ta	73	180.9479
Technetium	Tc	43	(98)
Tellurium	Te	52	127.60†
Terbium	Tb	65	158.9254
Thallium	Tl	81	204.383
Thorium	Th	90	232.0381
Thulium	Tm	69	168.9342
Tin	Sn	50	118.710⌐
Titanium	Ti	22	47.88†
Tungsten (Wolfram)	W	74	183.85†
Unnilhexium§	Unh§	106	(263)
Unnilpentium§	Unp§	105	(262)
Unnilquadium§	Unq§	104	(261)
Uranium	U	92	238.0289
Vanadium	V	23	50.9415
Xenon	Xe	54	131.29†
Ytterbium	Yb	70	173.04†
Yttrium	Y	39	88.9059
Zinc	Zn	30	65.39†
Zirconium	Zr	40	91.224

§ The International Union of Pure and Applied Chemistry (IUPAC) has not adopted official names or symbols for these elements.

Atomic weights corrected to conform to the most recent values of the Commission on Atomic Weights. Column nomenclature conforms to IUPAC system and data in this chart have been checked by the National Bureau of Standards' Office of Standard Reference Data.

⌐ These weights are considered reliable in the last place, as follows:
Calcium and Gallium ± 4; Boron ± 5; Chromium and Sulfur ± 6; Hydrogen and Tin -7
All other weights are reliable to ± 1 in the last place. All reliabilities are based on an uncertainty scale of ± 1 to 9.

* These weights are considered reliable to ± 2 in the last place.
† These weights are considered reliable to ± 3 in the last place.

c Copyright 1984
Fisher Scientific

Fisher Scientific